Science Firsts

FROM THE CREATION OF SCIENCE
TO THE SCIENCE OF CREATION

Robert Adler

John Wiley & Sons, Inc.

The author gratefully acknowledges the following sources for permission to use photographs in this book: Edgar Fahs Smith Collection, University of Pennsylvania Library (pp. 39, 47, 54, 57, 63, 70, 76, 88, 98, 103, 110, 116, 123, 127, 137); The Alfred Wegener Institute for Polar and Marine Research (p. 142); Dart Collection, University of Witwatersrand, Johannesburg, South Africa (p. 151); Fermi National Accelerator Laboratory (p. 159); Lucent Technologies, Inc./Bell Labs (pp. 170, 183); Jeremy Norman and the Archive for the History of Molecular Biology (p. 174); Paul Schnaittacher, courtesy of Lynn Margulis (p. 189); University of Geneva, Press Information Publications (p. 195); The Roslin Institute (pp. 202, 204); Roddy Field, courtesy of The Roslin Institute (p. 205).

Published by John Wiley & Sons, Inc., Hoboken, New Jersey
Published simultaneously in Canada

For general information about our other products and services, please contact our Customer Care Department within the United States at (800) 762-2974, outside the United States at (317) 572-3993 or fax (317) 572-4002.

Wiley also publishes its books in a variety of electronic formats. Some content that appears in print may not be available in electronic books.

ISBN 0-471-40174-9

Printed in the United States of America

10 9 8 7 6 5 4 3 2 1

To Jo

Your sparkling eyes,
mischievous spirit,
and mysterious soul
make every day with you
a discovery.

Contents

Acknowledgments

In a beautiful poem, e. e. cummings wrote, "My father moved through . . . haves of give." I think of that poem frequently. My father, Hy Adler, was an immensely loving, generous, and giving man. He helped me in countless ways throughout my life and at every stage of this book. I had the opportunity to thank him many times in person before his death at the age of 89, but I want to express my undying gratitude to him publicly as well. His vitality, charm, and creativity warmed and illuminated every moment with him. He will always be loved and missed.

I also want to thank Dr. Robert Utter, whose exuberant appreciation of knowledge and ideas, and whose willingness to search through and share his superb collection of books and articles on the history of science contributed greatly to *Science Firsts*.

Another person who helped greatly with my research is Jack Ritchie, Circulation Supervisor at the Sonoma State University library. I hope he will accept my thanks for making every visit to the library a pleasure. I am also grateful to John Pollack of the Annenberg Rare Book and Manuscript Library, who greatly facilitated my search for illustrations.

Thanks go also to my brother, Les Adler, my great friend, Lou Miller, and to the other members of the institute, for being unfailing sources of encouragement, ideas, and fun.

The person who has done the most to help bring this book into the world is my wife, Jo Ann Wexler. She has patiently given me enormous amounts of time and every kind of support. She has single-handedly kept our lives running smoothly, despite my long preoccupation with this project. It will take more than thanks to show my appreciation for all that she has done.

And of course, I want to thank my editor, Jeff Golick, for shepherding *Science Firsts* through from concept to completion.

Introduction

The most beautiful thing we can experience is the Mysterious.
It is the source of all true art and true science.

—*Albert Einstein*

A sense of mystery. Einstein had it, and so did Aristotle and Aristarchus, Curie and Dart, Margulis and McClintock. They never lost their childlike wonder about the universe, or, as Noam Chomsky described it, *"the ability to be puzzled by simple things."* Throughout history, a few people have combined that insistent tickle of curiosity with other qualities of mind and character that goad them into uncharted territories, stir them to ask probing questions, and attune them to new answers. Among those explorers, a very few are privileged to break new ground, discover new worlds, or glimpse shining new vistas no one has seen before.

The greatest of those—creators like Thales, Newton, Darwin, and Einstein—gave us radically new ways to understand the universe and our place in it. Like the great impact that ended the era of the dinosaurs 65 million years ago, their discoveries punctuated history, brought epochs to an end, and cleared the way for new systems to emerge. This book tells the stories of some of those gifted, driven, and complex explorers and the new worlds they were the first to enter.

Science Firsts starts in Greece 2,600 years ago. The sweeping curiosity of Ionian philosophers such as Thales, Anaximander, and Leucippus led them to ask basic questions about nature. How did the world come about? What is it made of? How does it work? Equally important, they were the first to insist on answers from within nature itself, to demand explanations that did not depend on the whims of the gods. By the time of Archimedes four centuries later, the Greeks had built the foundations of science. But the ancient world soon crumbled. Along with its theaters and temples, Greek science lay buried and forgotten—at least in the West—for a thousand years. Luckily, it was preserved, and in a few cases improved upon, by scholars in the Islamic world. Rediscovered, it helped ignite the Renaissance. That brings us to figures like Copernicus, Galileo, and Kepler, the heroic forebears of modern science. We have to peer across a thousand-year gap, but we can trace today's science straight back to Thales.

That's not to say that science only developed in the West. The precursors to science—the keen observation of nature, classifying, counting, remembering, discovering regularities—date back at least to the Neolithic. The people who reached and colonized Australia 60,000 years ago, who left us vivid paintings of cave bears and shamans in France and Spain 30,000 years ago, also left us cryptic signs that they counted the months, tracked the phases of the moon and the changing seasons, and studied the skies. Ethnobotanists working today praise the sophisticated biological knowledge of indigenous groups, and exploit it to identify medicinal plants and substances. The very survival of our ancient ancestors suggests that they studied their environments just as avidly. Led by Joseph Needham, scholars have discovered that technology, mathematics, and science all have deep roots in China. Many significant inventions and innovations appeared in China centuries or even millennia before they made it to the West. India, too, can claim its own early and important contributions, especially in physics, mathematics, and medicine. And the Greeks benefited from centuries of work in astronomy, mathematics, and medicine by anonymous Egyptian and Babylonian scholars.

Science Firsts, however, will follow the stream of Western science. The story starts in the world of Homer and Hesiod, where the stormy skies and the wine-dark sea were ruled by the gods, and where the origins of things were revealed—or obscured—in myths. Each bit of understanding that we've gained since then has been won by someone pushing into the darkness, breaking free from preconceptions, and glimpsing something clearly for the first time. Most of those steps must have been small and anonymous, and many must have led nowhere. But a few were brilliant and beautiful. As Gustave Flaubert wrote, "Amongst those who go to sea there are the navigators who discover new worlds, adding continents to the Earth and stars to the heavens: they are the masters, the great, the eternally splendid." It's those navigators we'll follow, using their hard-won flashes of insight to light our way.

1

Thales and Natural Causation

Blessed is he who has learned how to engage in inquiry,
with no impulse to harm his countrymen or to pursue wrongful actions,
but perceives the order of immortal and ageless nature, how it is structured.

—*Euripides*

In the beginning was a question. Twenty-six centuries ago Thales (*c.* 624 B.C.—*c.* 547 B.C.), a citizen of the Greek colony of Miletus, asked, "What is the world made of?" With that question, and by his insistence that it not be answered by a story about the gods, Thales planted the seed that would grow into Western science. The answer he gave was unsatisfactory, as his pupil Anaximander soon pointed out. Yet, by basing his speculations on observation and reason rather than revelation, Thales invited others to criticize his ideas and offer arguments and answers of their own. Anaximander was the first to accept that invitation. The dialogue they began—marked by an open clash of competing ideas, and with the ultimate appeal not to the whims of the gods but to nature and reason—marks the birth of science.

We know little about Thales. Historians conjure up the dates of his birth and death from events that took place during his life. He was born in Miletus, then a bustling regional center on the southern coast of what is now Turkey. His father was a Carian named Examyes, his mother, Cleobuline, was probably Greek. Miletus's cultural roots lay in mainland Greece, but it thrived because of trade throughout Asia

3

Minor and the Middle East. Thales appears to have been a man of affairs. He seems to have traveled widely, and is credited with being the first to bring geometry and astronomy to Greece from already ancient Egypt. He is reputed to have been an engineer who was able to change the course of a river so that an army could cross it. The later Greeks listed Thales as one of their Seven Sages. Like those other revered figures, his expertise included politics. The farsighted Thales warned the Ionians that they needed to unite to defend themselves against the Persians. Ionia remained divided and fell to the Persians fifty years after Thales's death.

Plato and Aristotle tell two very different stories about Thales. With a suspicious wealth of detail, Plato writes, "Theodorus, a witty and attractive Thracian servant girl, is said to have mocked Thales for falling into a well while he was observing the stars...." That would make Thales both the first philosopher and the first absent-minded one. In contrast, Aristotle tells us that, based on his knowledge of astronomy, Thales was able to predict a bumper crop of olives and parlay that prediction into a fortune by gaining control of the region's oil presses. That would make him the world's first scientist-entrepreneur.

Absent-minded or shrewd—we can't be sure, but we do know that Thales was the first person we can name who asked a fundamental question about nature and answered it on strictly natural terms. Long before anyone had thought of atoms, before there were words for matter, science, or even philosophy, Thales sought to know what the world was made of. He refused to rest with how things seemed. He believed that mountains and seas, plants and animals, wind and rain—all the things we perceive—stem from a common source. And, crucially, he was not willing to accept an answer invoking the gods or anything else above or apart from nature. An angry Zeus was not the source of thunder and lightning, nor was "broad-bosomed Earth" created by Chaos. Thales was a politician and engineer. To his practical mind, everything must have evolved or differentiated from something real, something he could see and touch.

Aristotle restated Thales's great insight two and a half centuries later: "For there has to be some natural substance, either one or more than one, from which the other things come to be, while it is preserved." The choice Thales made for that primordial substance was water. We don't know why. Aristotle speculated that Thales looked at creation biologically, observing that all living things contain water, and that the processes of insemination and nutrition involve moisture.

But, as we'll see, Aristotle was fascinated by living things, a fascination there is no evidence that Thales shared. It's equally likely that Thales observed the different physical states of water—solid, liquid, and mist or vapor, and reasoned that this protean substance could account for all the varied things of the world.

As bold as he was, Thales could not divorce himself totally from ancient traditions. He found it difficult to explain the movement of wind and water, lodestones and living things, on the basis of substance alone. But he also refused to look for the source of nature's dynamism outside of nature. Instead, he invested everything with a kind of life force. Later philosophers would call Thales and his followers *hylozoists*—those who believe everything is alive.

Thales went on to create a model of the universe—the first strictly physical cosmology. Earth, he thought, had formed from the primordial waters, like the Egyptian delta emerging from the Nile. He conjectured that the Earth was a flat disk, floating like a log. Earthquakes, sensibly, were caused by waves in the surrounding waters. The heavens were circled by a great river, with the sun, moon, planets, and stars being blown across the sky by winds stirred by the water's circulation.

Anaximander, Thales's brilliant pupil, soon devised a much more sophisticated picture of the cosmos. But he addressed the same kind of questions Thales asked. What is the world made of? How did it develop? What keeps the Earth in place? In turn, Anaximander's student Anaximenes criticized both his predecessors and developed his own models and explanations. However much these first philosophers differed, they shared two beliefs: nature must be understood without resorting to supernatural causes, and humans are capable of discovering nature's truths through observation and reason.

As if it were not enough to have invented scientific inquiry, Thales was also celebrated as an astronomer. The most hotly debated of his feats is his supposed prediction of a solar eclipse that marked the end of a long war between the Lydians and the Medes. Both ancient and modern writers rightfully saw this as a remarkable accomplishment—one that other astronomers would not be able to duplicate for centuries. The Roman historian Herodotus repeated the story, which he derived from earlier sources. The eclipse in question must have occurred during the Forty-ninth or Fiftieth Olympiad (between 585 and 577 B.C.). In the nineteenth century, astronomers calculated that a total eclipse of the sun had in fact darkened the skies of Ionia on May 28, 585 B.C. That, they believed, was Thales's eclipse.

Until recently, most historians accepted the story. They assumed that Thales had gleaned astronomical knowledge from the Egyptians that enabled him to foretell the time and place of the eclipse. Scholars knew, for example, that in the second century A.D. the great astronomer Ptolemy studied Babylonian eclipse records dating back to 747 B.C. Today, however, with much greater knowledge of ancient astronomy, historians of science are convinced that the best the Babylonians or Egyptians could do was to identify periods when solar eclipses were possible somewhere on Earth. However good their records, they could not predict that a solar eclipse would definitely take place, much less pin it down to a particular locale. In light of the olive-press story, it's not difficult to imagine that Thales might have been enough of a risk-taker to predict an eclipse even if he was far from sure of being right. Still, scholars today suggest we reverse our understanding of the tale: Thales didn't become famous because he foretold an eclipse. Rather, later writers attributed the prediction to him because of his fame. It's almost certainly a myth, like the story of George Washington throwing a silver dollar across the Potomac.

Putting the stories aside, Thales still stands as a heroic figure, a true culture-giver. Even if all he did was to see past the bewildering variety of what we perceive, to ask what all things are made of, he would deserve his fame. But by insisting that the answer must be found within nature, not above it, he gave us our first scientific tools. The classical scholar G. E. R. Lloyd credits Thales with nothing less than "the discovery of nature."

Far more remarkable than any story about Thales is the fact that twenty-first-century physicists, wielding the most powerful experimental tools ever devised, are simply attempting to complete what Thales started 2,600 years ago. By blasting atoms together with enormous energy they are recreating the conditions that existed an instant after creation—a time when all the forces of nature were unified and matter was reduced to its fundamental constituents. Aristotle described the goal that Thales and all his followers sought:

> that from which all things are, and out of which all things come to be in the first place, and into which they are destroyed in the end—while the substance persists, but the qualities change—this, they say, is the element and first principle of things.

2

Anaximander
Orders the Cosmos

Anaximander of Miletus, the pupil of Thales, was the first to depict the inhabited Earth on a chart. After him Hecataeus of Miletus, a much traveled man, made it more precise so as to be a thing of wonder.

—The geographer Agathemerus

What the system of Anaximander represents for us is nothing less than the advent, in the West at any rate, of a rational outlook on the natural world.

—Charles H. Kahn

A naximander (*c.* 610 B.C.—*c.* 546 B.C.) was an incredibly bold thinker. With great expansiveness of mind he asked, and answered, one of the prototypical questions of early Greek science—how did the world come to be? Realizing that the substances and qualities we perceive inevitably change and pass away, he postulated the existence of the *apeiron*, "the boundless." The *apeiron* was material, but with no beginning or end in time or space. It served as both the source and fate of everything we see. The enormity of that concept forced him to drastically revise the status of the Earth. To Anaximander the Earth, in fact our whole cosmos, is not only finite in size and limited in duration, but is just one of an infinite number of worlds. What an amazing—and chilling—degree of objectivity to be achieved more than 2,500 years ago.

Who was this remarkably imaginative man? Like his mentor, Thales, we know little about him. Anaximander, son of Praxiades, was born in Miletus. One source tells us he was sixty-four years old around

7

546 B.C. He is said to have led a political delegation to Sparta, where he presented the Spartans with two of his great innovations—a sundial and his map of the world. He may have founded a new Milesian colony in Apollonia, near the Black Sea. There's a tradition that he was something of a showman, dressing and speaking dramatically. He was the first philosopher to write his ideas down in prose rather than in the poetic tradition of Homer or Hesiod. It's a measure of how long ago he lived, and how many layers of history separate us from him, that only one cryptic sentence survives of all he wrote:

> That from which all things are born is also the cause of their coming to an end, as is meet, for they pay reparations and atonement unto each other for their mutual injustice in the order of time.

Anaximander created the first coherent naturalistic system of the world. He believed that a primordial undifferentiated substance, "the Boundless," has always existed and is always in motion. Just as flowing water can spontaneously spawn a whirlpool, the boundless spontaneously generated the rotating germ or seed of the world. Once formed, the qualities of hot and cold, and later of dry and wet, separated out and began to interact. The hottest material moved outward, leaving a cool, wet interior surrounded by a fiery shell. The intense heat caused moisture to evaporate, building up pressure and eventually blasting the shell apart. Its remnants coalesced into rotating rings of fire surrounded by opaque tubes of mist. Eventually the heat evaporated enough water to expose dry land, creating the Earth on which life eventually emerged. Light blazing out through openings in the fiery tubes appears to us as the Sun, the Moon, and the stars. Rhythmic changes in the openings cause the phases of the Moon as well as eclipses of the Sun and Moon.

One of Anaximander's remarkable accomplishments was to propose an evolutionary theory twenty-three centuries before Darwin. Anaximander argued that all land animals, including humans, evolved from fishlike ancestors. He thought that the earliest forms of life arose spontaneously through the interaction of primordial warmth and moisture. Those first creatures, protected by barklike shells, lived in the seas. As dry land appeared, some were faced with the problem of adapting to new conditions. Again Anaximander shows the remarkable extent to which he was able to free himself from any trace of anthropocentrism. Humans evolved from water creatures just as did all other land animals, with one difference. Because human infants are so helpless at birth, he surmised that they must have been nurtured by some other kind of sea creature before they could survive on land.

Clearly, Anaximander did not create a full-blown theory of evolution capable of explaining the descent of all creatures. That would be left for Darwin to accomplish. Still, Anaximander caught at least a glimpse of evolution. As in Darwin's day and our own, it's an idea that many people found disturbing. Writing 150 years after Anaximander, Plato chose to believe in the transmigration of souls, first taught by Pythagoras. Plato turned the idea into a kind of reverse evolution, arguing that immoral or stupid men were reborn as animals (or as women). It's not difficult to guess whom Plato was skewering when he said, "The fourth kind of animal, whose habitat is water, came from the most utterly mindless men."

Mindless or not, Anaximander understood and utilized one of the basic assumptions of science — that the same processes that take place on Earth must occur throughout the universe. This belief led Newton, twenty-two centuries later, to propose the law of universal gravitation, which explains both the fall of an apple and the orbit of the moon. Anaximander realized that the same sequence of events that spawned the Earth and the bodies around it must occur over and over within the boundless, at many other places and times. So he boldly proclaimed that there must be an infinite number of worlds that, like our own, are born, exist for a time, and, "paying reparations," cycle back into nothingness.

Anaximander's conceptualization of the boundless allowed him to solve a problem that had bedeviled his mentor, Thales, and would stymie his successor, Anaximenes. In their attempt to give a satisfactory answer to the question of what kept the Earth in place, Thales had it floating on water and Anaximenes saw it cushioned on air, but neither of them explained what held up the supporting substance. Anaximander, with his characteristic incisiveness, divorced himself completely from the earthbound sensory evidence of up and down. He envisioned the Earth at rest within an infinite, symmetrical universe. Why should it fall? It was in equilibrium, "not dominated by anything," and therefore had no more reason to move in one direction than another. In this, as in many other of his ideas, he was far ahead of his time.

Once he had freed the Earth from its supports and could visualize it floating freely within the cosmos, Anaximander could also surmise that it had another, potentially habitable side. He pictured the Earth as a short, drum-shaped cylinder. The flat surfaces, he thought, are three

times larger in diameter than the distance between them. The inhabited world forms one of the flat surfaces. We don't know if he believed
that people lived on the opposite surface. However, someone who
could envision an infinite number of worlds almost certainly would
not have balked at picturing humans on the other side of the Earth.

Anaximander reduced the Earth to an infinitesimal dot in a boundless universe, but he also made it a worthy object of study in its own
right. Later geographers identified him as the first to draw a map of the
world. Since there are surviving examples of extremely simple world
maps from Babylonia from about the same time, he may well have been
inspired by examples he had seen or heard described. His map has not
survived, but it most likely showed the inhabited surface of the Earth as
a circle completely ringed by an ocean. The known landmasses may
have been shown surrounding what we now know as the Mediterranean
Sea. It probably stretched as far west as the Pillars of Hercules—our
Gibraltar—east to Babylonia, north into Europe, and south into Libya.
Inspired by Anaximander, mapmaking progressed rapidly. A few decades later another Milesian, Hecataeus, would improve Anaximander's
map based in part on his own extensive travels, making it "a thing of
wonder." Writing just a century after Anaximander's death, the historian
Herodotus found the old circular maps laughably out of date.

Anaximander's interest in mapping was not limited to the Earth.
He is also said to have produced a map of the heavens in the form of a
sphere. He appears to have divided up the heavenly sphere into bands,
some of which crossed the others. This may be why he is said to have
discovered the obliquity of the ecliptic. We don't know the details, but
if in fact he mapped the heavens as a sphere, it would have been an
extremely significant step. The Babylonians and Egyptians, with
whom the ancient Greeks had contact, had accumulated centuries of
astronomical observations. They had noticed patterns within them,
which they used to keep their calendars in order and to make predictions. But they had not come up with a physical model of the heavens.
Anaximander may have been the first to wed Eastern astronomy to
Greek geometry, in the process creating a powerful unifying system.

The details of Anaximander's cosmology, like his sketchy map,
may now seem laughable. But by envisioning the cosmos as an understandable whole, and by insisting that its origin, development, observ-
able phenomena, and fate can all be explained as the dynamic
interaction of basic and universal laws "in the order of time," he set in
place the conceptual foundation of science.

3

Pythagoras Numbers the Cosmos

All things are number.

—*Pythagoras*

And indeed all the things that are known have number.
For it is not possible that anything whatsoever be understood
or known without them.

—*Philolaus of Croton*

Both the unity and structure of the whole world and the specific
nature of each thing are expressed by simple numerical ratios, and
this is what makes them knowable. This is as far as we can go in
recapturing the central doctrine of Pythagorean philosophy.

—*Charles H. Kahn*

Trying to understand Pythagoras (*c.* 570 B.C.—*c.* 490 B.C.) is like studying the remains of an ancient supernova. All that we can see today is an expanding cloud of incandescent gas and obscuring dust. We infer that only a titanic explosion could have produced this phenomenon. But when we try to look into the center to make out the explosion's source, the very brilliance of the fireball makes it impossible.

What we do know is that around the year 530 B.C. Pythagoras, son of Mnesarchus, left his native Samos in Ionia and traveled to Croton in what is now southern Italy. He was forty to fifty years old at the time. Earlier, he is said to have consulted with the aging Anaximander, and on his advice lived and studied in Egypt and Babylonia for many years. Historians say that all we know for sure about Pythagoras is that he founded a religious sect that quickly spread from Croton throughout southern Italy, becoming more controversial as its influence grew. The Brotherhood was a mystical sect, and Pythagoras was its unquestioned leader. He taught that the human soul is immortal, imprisoned in the body, and that on its journey toward perfection the soul is reborn within people or animals. He attracted hundreds of disciples who gave up their possessions, lived communally and simply, and tried to purify themselves through carefully prescribed practices in order to grasp the mysteries the Master taught.

What distinguishes the Pythagoreans from hundreds of other cults throughout history is that the core of their beliefs, the object of their devotion, was number. To Pythagoras and his followers, numbers were divine; they saw in them nothing less than the ultimate source and organizing principle of the cosmos. Today, with 2,500 years to insulate us, we can coolly say that Pythagoras shifted the focus of scientific thought from matter to form. His Ionian predecessors—Thales, Anaximander, and Anaximenes—had tried to understand the universe in terms of substance. Pythagoras's searing insight, his encounter with the divine, was that without form, pattern, organization, relationship—most purely expressed in numbers—nothing could exist. Anaximander's Boundless, or Unlimited, required a limiting principle in order to differentiate into a structured whole, a cosmos. By emphasizing form, Pythagoras began an essential dialogue that continues to pervade science today. One example is quantum theory, which replaces entities with probabilities, substance blurs into pure information.

We may never know if it was Pythagoras or one of his disciples who discovered the elegant relationship between the length of a plucked string and the musical note it produces. For the first time a subjective human experience—the harmonies we hear when chords such as the octave, fifth and fourth are played—was shown to have a purely mathematical basis. The substance of the vibrating string didn't matter; the numerical relationships did. Lengths in the ratio of 2:1 always produced an octave, 3:2 a fifth, and 4:3 a fourth. "It was the first successful reduction of quality to quantity," writes Arthur Koestler, "and therefore the beginning of Science."

It was a magical discovery to Pythagoras and his followers, as surprising as if the stars had arranged themselves into legible words across the sky. They found the metaphor for creation in those humming strings and integral ratios. The unbounded range of pitch paralleled the primordial Unlimited. Number was the limiting principle that created notes and the harmonious relationships between them, that created form and pattern and determined what humans saw and heard.

It was in Pythagoras's fiery mind that this observation exploded into a universal law. If the first four numbers governed sound and music, then numbers and their relationships must be everywhere. More than that, numbers reigned—substance was not important, as long as it could be divided according to these divine ratios. To Pythagoras, number was the eternal First Principle sought by his Ionian predecessors.

To the Pythagoreans, numbers possessed an even more remarkable power. Numbers, they believed, are not just abstract ideas, not simply the source of order, but comprise the actual stuff of the universe. The Pythagoreans turned numbers into "figures" by forming dots into lines, lines into two-dimensional figures such as triangles, and those into three-dimensional solids such as pyramids. They envisioned a kind of numerical atomism—long before material atoms were dreamed of—in which numbers actually built lines, lines built planes, and planes built solid objects. The philosopher Parmenides, and later Zeno with his famous paradoxes, subjected this materialization of numbers to scathing criticism.

Pythagoras and his followers elaborated the primacy of numbers in a variety of areas. They created a cosmogony, or theory of origins, in which the primordial Unlimited was breathed in by the Limiting— the "inspiration" that produced our ordered universe. Pairs of opposite qualities differentiated next. This led to a symmetrical, finite cosmos in which the spherical heavens rotated serenely, although not around either the Earth or the Sun. To the Pythagoreans, the Earth was not pure enough to occupy the center of the cosmos. At the center, and beyond the sphere of stars as well, was fire. The Earth, Moon, planets, and stars all circled the Central Fire. This unprecedented decentering of the Earth proved to be a disturbing and unacceptable idea. It would lie dormant in the European unconscious for two millennia, until it whispered in the ear of the young Copernicus.

Pythagoras saw the same pattern in the structure of the cosmos that he had discovered in music. Number imposed structure and order on the limitless void. As if the strings of a lyre had been formed into perfect circles around the sky, the planets and stars circled in properly proportioned, well-tempered orbits. Their varied movements created a

chorus of sound, the "harmony of the spheres." It was divinely beautiful, his followers believed, but sadly not for mortal ears to hear.

It is just possible that, inspired by the perfection of the sphere, and
having made Earth a planet circling the central fire, Pythagoras may
have been the first to teach that the Earth was spherical. Plato, writing a
century after Pythagoras died, was the first to mention the idea. He
linked the idea to the Pythagoreans. Later writers ascribed the idea
either to Pythagoras himself or to Parmenides, first a follower and later a
cogent critic of Pythagorean beliefs. However, most modern scholars
say that too few clues remain; the case simply can't be proved. We may
never be sure if it was Pythagoras who first perceived Earth's true figure.

The Pythagorean glorification of numbers could easily have led to
futile numerology. In fact, much of their use of numbers as symbols
seems arbitrary today: the number four, two times two, represented balance, hence justice. Two was female, three male, so five represented
marriage. Yet despite their mystical ideas, the Pythagoreans made great
progress in mathematics, especially geometry and the theory of numbers. Again, Pythagoras's own discoveries cannot easily be sorted out
from the work of the mathematicians he inspired. But it is clear that
Pythagoras founded number theory, and that he and his followers discovered many of the geometrical theorems and methods that Euclid
systematized in his *Elements*.

It is now equally clear that Pythagoras did not discover the famous
theorem named after him—that the sum of the squares on the sides of
a right triangle equals the square of the hypotenuse. The historian of
science Otto Neugebauer has deciphered Babylonian texts showing
that mathematicians there grasped this idea a thousand years before
the Greeks.

The Brotherhood Pythagoras created was shattered near the end
of his life. It had become a powerful political force throughout southern Italy, which provoked an angry reaction. Meetinghouses were
burned; Pythagoras and his followers had to flee or be killed. Pythagoras had not recorded any of his ideas; he communicated his insights
face-to-face, in bits and pieces, symbols and aphorisms. "Don't stir the
fire with a sword," he said, or, "Don't step beyond the center of the
balance." Only his inner circle, the esoterics, could speak directly with
him. They too wrote nothing down. Philolaus, born a century after
Pythagoras, was the first to write out a version of Pythagorean philosophy. Unfortunately, neither he nor later Pythagoreans told us which

ideas came directly from the Master, which had developed under his watchful eye, and which appeared after his death.

Despite the dramatic end of his Brotherhood, Pythagoras and his followers indelibly influenced Greek philosophy, mathematics, and science. We even owe our word for mathematics to him and his followers. Plato absorbed and transformed many Pythagorean concepts, including the primacy of ideas in general and of mathematics in particular. Through Plato, Pythagoras shaped Western thought. His ideas found fertile soil in the Renaissance. From Galileo on, the language and primary tool of science has been mathematics. More generally, Pythagorean ideas of symmetry, balance, and harmony intertwine through the foundations of modern science, medicine, and government.

Shakespeare, in *The Merchant of Venice*, gives voice to the Pythagorean harmony of the spheres. Pythagoras was the first to hear that cosmic music. It would be 2,200 years before another strange and inspired soul, Johannes Kepler, would hear it too.

> Look how the floor of heaven
> Is thick inlaid with patines of bright gold;
> There's not the smallest orb which thou behold'st
> But in his motion like an angel sings
> Still quiring to the young-eyed cherubins.
> Such harmony is in immortal souls;
> But whilst this muddy vesture of decay
> Doth grossly close it in, we cannot hear it.
>
> —*The Merchant of Venice*, Act V, Scene 1

4

Atoms and the Void

Nothing happens by chance, but everything from reason and by necessity.

—*Leucippus*

By convention there is sweet, by convention bitter, by convention hot, by convention cold, by convention color; but in truth there are atoms and the void.

—*Democritus*

More than 2,400 years ago, on the shores of the Aegean Sea, a philosopher thinking about the nature of things came upon one of the most fertile ideas of all time—that everything is made of atoms moving through space. We know little about Leucippus as a person, but there is little doubt that he discovered the atom and gave it its name.

Leucippus defined atoms as minute, indivisible particles of pure substance, eternally in motion. They were hard, compact, and homogeneous. They had no parts. They took up space, but contained no empty space. They could bounce off or hook onto each other, but were perfectly rigid—their size and shape could not be changed. The objects they combined to form might come and go, but the atoms were indestructible. They had always existed and always would. Leucippus used the Greek word *atoma*—uncuttable—to describe them.

Leucippus was born in Miletus, the same city that had produced the first natural philosophers, Thales, Anaximander, and Anaximenes. As an adult, Leucippus may have moved to Abdera, on the north coast of the Aegean, where he taught his new atomic doctrine. It's likely that he wrote at least one work, "The Greater World-System," now lost. Abdera was the hometown of Democritus, Leucippus's most brilliant disciple. In fact, Democritus would became so famous that Epicurus, who wrote on atomism a century later, denied that Leucippus had ever existed.

Democritus seems to have been blessed with wealthy parents. On his father's death, he spent his inheritance on travels, which appear to have been extensive. Among his seventy-three published works, all of which have also been lost, were treatises on Babylonia, Chaldea, and a voyage "around the ocean." Democritus wrote about mathematics, physics, geography, medicine, and most other subjects known to the Greeks. His lasting fame flows from his elaboration of the atomic theory, but in Greek and Roman times he was at least as famous for his observations on ethics and government.

A story about Democritus illustrates both his legendary kindness and his belief in his atomic theory. Having gone blind and in failing health, he decided to end his life by refusing food. However, so as not to spoil his sister's enjoyment of an important festival, he postponed his death—not by eating, but by inhaling the sustaining atoms he believed wafted from freshly baked bread. His range of knowledge earned him the nickname of "Sophia," or Wisdom. His wry perspective on human foibles and his prescription for a good life—cheerfulness—earned him a second nickname, "The Laughing Philosopher."

Leucippus did not develop the atomic theory to explain any specific observations or physical phenomena. Rather, he proposed the atom as a radically new answer to the question Thales had asked 150 years earlier—what is the ultimate nature of things? By the time of Leucippus, logicians such as Parmenides and Zeno had honed philosophical dialogue to a razor's edge. They used incisive questions and paradoxes to critique the work of earlier philosophers. If there is one substance or principle, how can many things exist? How can we account for the apparent permanence of objects, and for coming into being, motion and change? Can matter be divided infinitely? If so, how does it produce extension in space? Can space be divided infinitely? If so, how can there be movement? Can time be divided infinitely? If so, how can there be change? It was these philosophical questions, not a physical observation or problem, that Leucippus answered with a pair of immensely productive concepts: atoms and the void.

To Leucippus, the infinite or boundless void was just as fundamental as atoms. "The elements are the full and the empty," he said. Wherever there were atoms there was no void, and wherever there was void there were no atoms. Together they comprised the universe. Leucippus simply assumed the existence of the void as the necessary complement to his material atoms. The void, he argued, was no more or less real than atoms. The void prevented atoms from merging, keeping them apart no matter how close they came to one another. And since the void was everywhere, his ever-moving atoms always had somewhere to go.

Leucippus kept his atoms as simple as possible. They were all made of the same primary substance, but varied in three ways, which he called, "rhythm," "touching," and "turning." "Rhythm" seems to have encompassed both size and shape. Somewhat like Lego blocks (and very much like protein molecules), Leucippus's atoms came in a wide variety of forms. Small, spherical atoms were the neutrinos of his system, able to zip through matter with ease. Larger, more irregularly shaped atoms had greater combining power—they tended to get tangled up with each other to form sensible objects such as rocks, the Earth, and human bodies. "Touching" described such atoms in close relationship to each other—although he made it clear that they were always separated by at least a bit of void, and so could never merge. "Turning" described their orientation in space.

Leucippus and Democritus used the analogy of letters and words to illustrate how atoms differed and how they joined to form compounds and objects. For example, the letters "A" and "N" differ in rhythm, the letters "N" and "Z" differ in turning, and the sequences "ON" and "NO" differed in touching. To emphasize how much variety different combinations of atoms could create, Democritus observed that switching just one letter in a play might change it from a tragedy to a comedy.

Leucippus, the pioneer, and Democritus, his disciple, had enormous faith in their atomic system. So much so that they were willing to reduce all causation to the natural, mechanical actions and interactions of atoms. They jettisoned Thales's life force, Anaximander's warring opposites, Empedocles's "Love and Strife," and Anaxagoras's "mind." What remained was "reason and necessity"—the logic of natural laws governing interacting atoms. Atoms were always in motion. When they collided, they might rebound or connect. The only "force" they felt was a tendency, like flocking birds or pebbles on a beach, for similar forms to associate.

Democritus pushed the mechanistic model to the limit. He argued, as Newton would two millennia later, that the entire course of the universe was foreordained by the initial configuration of its particles and implacable natural laws. Democritus applied this not only to inanimate objects, but also to living things, to sensation, thought, and to the soul. His stark view of causation provoked a great deal of criticism, especially from Aristotle, who firmly believed that events unfold toward a "final cause" or goal. To Leucippus and Democritus, nature evolved by blind necessity, not by intelligence or design.

In the infinity of the void, however, blind necessity could easily create a cosmos. For Leucippus, it only took a "whirl" to create a world. He believed that the random motions of large numbers of atoms within the void would inevitably form a vortex, a whirlpool in space. Within the whirl, atoms would sort themselves out not by weight but by size and shape. Larger and more irregularly shaped atoms would accumulate in the center of the vortex, squeezing out the smaller, more mobile atoms. A spherical membrane or shell would form, and within it an earth. Both the Earth and its surrounding membrane started off moist. As the Earth dried, it revealed landmasses surrounded by oceans. As the rotating sphere dried, parts of it caught fire, becoming the Sun, Moon, and other luminous bodies. In many ways it's a remarkable preview of current models of the formation of stars and planets from rotating clouds of dust and gas.

Nor did Leucippus or Democritus think that our world was unique. They were sure that an infinite number of vortices would form, creating an infinite number of worlds. Democritus had no trouble imagining these worlds. Some would have no sun or moon, others might contain even larger bodies, or more of them. Some might be moist like Earth, and support living things, others dry and sterile. At any given time, some worlds are forming, others are in their prime, and others are being destroyed. Worlds could collide, destroying or cannibalizing each other. It's a tribute to the intellectual freedom of the Greek world that Democritus could advance such ideas and risk only intellectual criticism. Nearly two thousand years later, Renaissance philosopher Giordano Bruno would echo him, writing, "Innumerable suns exist; innumerable earths revolve around these suns. . . . Living beings inhabit these worlds." For such heresies, Bruno was condemned by the Inquisition and burned at the stake in Rome.

Democritus didn't stop with the cosmos; he also worked out a mechanistic, atom-based theory of the psyche, or soul. He explained all life processes, including respiration, perception, and thinking in terms of atoms in motion. He thought that coordinated movements of especially fine, easily moved atoms, similar to those in fire, carried out these vital functions. Those mobile, sensitive atoms permeated the entire body, but clustered together, he thought, in the heart.

To explain perception, Democritus argued that something physical must connect an object and a sensory organ. This was easy to understand for touch and taste, but very difficult to explain for the other senses. Democritus came up with imaginative descriptions of how smells, sounds, and images were conveyed by atoms. Some of his schemes are less than convincing, and were much criticized by later writers. For example, he thought that thin sheets of atoms, which he called "idols," constantly peel away from the surface of objects and enter the eye. There they interact with soul-atoms to produce our subjective experiences. It is a bit far-fetched. Still, Democritus was the first to distinguish clearly between the primary properties of atoms—size, shape, position, and motion—and the subjective sensations, or secondary properties, we ascribe to things—qualities such as sweet or sour, red or green, loud or quiet.

Having worked out his mechanistic psychology, Democritus could address one of the other burning issues of Greek philosophy—what can we know? Given the tenuous, multistep connection between things and our perceptions of them, he could hardly argue that we can know the truth for sure. "We know nothing truly," he wrote, "for the truth lies hidden in the depth." Still, he had great faith in the ability of careful observation and reasoning to probe those depths and come close to objective reality. And that ultimate reality, he believed, was nothing other than "atoms and the void."

Democritus's insistence that the universe and everything in it evolve according to the predictable effects of physical laws on material bodies became the foundation of Western science. It reached its peak with Newton's perfectly regulated clockwork universe. The scientific view of atoms remained essentially unchanged until the nineteenth century, when chemists began to glean information about atoms and their combinations. It was only in the twentieth century that Leucippus's indivisible and imperishable atoms were found to mutate and divide.

Still, the fact that Leucippus and Democritus were able to infer the existence of atoms through nothing more than the exercise of reason, twenty-two centuries before scientists would uncover even the slightest chemical or physical evidence for them, stands as a stellar accomplishment of the human mind. "Atomism," writes science historian Lancelot Whyte, "has proved the power of the intellectual imagination to identify aspects of an objective truth deeply rooted in the nature of things." Perhaps Democritus had good reason for his optimism about the human ability to understand the cosmos.

5

Aristotle and the Birth of Biology

[W]hen we turn to the plants and animals that perish, we find
ourselves better able to come to a knowledge of them, for we are
inhabitants of the same Earth. Anyone who is willing to take the necessary trouble can learn a great deal about all the species that exist.

—*Aristotle*, The Parts of Animals

Plato called him "The Mind." His fellow students at the Academy
called him "The Reader"—not necessarily a compliment, since
philosophers were not supposed to dirty their hands, even by reading.
As an Academician he followed Plato for many years in arguing that
the world we perceive is just a crude image of the perfect and eternal
world of ideas. But in his late thirties, probably when he left Athens to
return to his native Macedonia, Aristotle (384 B.C.–322 B.C.) stopped
searching for Truth through philosophical disputation alone and instead sought a different kind of knowledge through the patient observation of hawks and honeybees, dolphins and dogfish, and in the
blood and stench of the dissecting table. There, face-to-face with nature's abundance and variety, he became the first to grope toward a
scientific classification and understanding of living things—not
through preconceived ideas, but on their own terms.

Aristotle's father, Nicomachus, was a physician at the court of
Amyntas II of Macedon. So, as a child, Aristotle may have been exposed to medicine as both a body of knowledge and a hands-on prac-
tice. Both his parents died when he was young, however, so he could

21

not have received more than initial training as a physician. At about the age of seventeen, Aristotle went to Athens to study at Plato's Academy. He stayed there for twenty years, evolving from a gifted student to a leading philosopher probing the nature of reality, knowledge, logic, and causality. It's significant that Aristotle was an Ionian. He seems to have absorbed the spirit of the great naturalistic philosophers who studied and taught there centuries before he was born. Thales and his followers had tried to explain the cosmos through natural causes rather than through myths, and had looked for the basis of things not in the whims of the gods, but within nature. The Athenian philosophers of Aristotle's day viewed the Ionian philosophers as naive, because they believed in the reality of what they saw rather than in the pure products of reason.

Aristotle left both the Academy and Athens in 340 B.C., at about the same time that Plato died. Anti-Macedonian feelings were being whipped up in Athens by the fiery orator, Demosthenes, leading Aristotle to stay away for twelve years. While in exile at the court of Hermias in Asia Minor, he married the ruler's niece Pythias. The couple and their daughter moved to the island of Lesbos, where Aristotle seems to have begun his studies of the animal kingdom, observing the creatures found on and around the island. He may have been assisted by Theophrastus, a native of the island, who would go on to complement Aristotle's study of animals with his own study and classification of plants. Famously, Aristotle spent three years tutoring Phillip of Macedon's son, soon to be known as Alexander the Great.

Aristotle was fifty years old when he returned to Athens, by then under the control of Alexander. He did not go back to the Academy, but instead founded his own school of philosophy, at the Lyceum. There he taught many subjects, but also continued his studies of animals. Aristotle eventually produced a series of books that form the foundation of biology. The most remarkable is his *Historia Animalium*, which touches on the physical structure and lifestyles of hundreds of kinds of animals—how they breed and reproduce, where they are found, and how they interact. He accompanied this with more specialized books including *De Partibus Animalium*, which compares the anatomy and functional physiology of animals, *De Motu Animalium* and *De Incessu Animalium*, dealing with how animals move, *De Generatione Animalium*, which traces how animals develop and grow, and *De Anima*, in which he explored what differentiates living and nonliving things.

As a biologist, Aristotle was not afraid to get his hands dirty. He was the first to systematically compile all he could from previous observers, not just philosophers, but fishermen, farmers, travelers, and other people with first-hand knowledge of animals. Even more importantly, he spent years patiently observing, studying, and dissecting animals. In all he described nearly six hundred species, including insects, crustaceans, fish, amphibians, reptiles, mammals, and birds. The quality of his observation is remarkable. He discovered, for example, that the blind mole has a hidden eye; that despite their male-like genitalia, female hyenas are not hermaphrodites; and that among river catfish, it's the male that takes care of the young. He studied honeybees in great detail, trying to puzzle out how they reproduce. In the process, he seems to have been the first, by 2,100 years, to notice their dance language.

It was in his writings on biology that Aristotle most clearly put observation above theory. After presenting his conclusions about how bees reproduce, he added, "But the facts have not been satisfactorily ascertained, and if ever they are, then credence must be given to observation rather than to theory, and to theory only in so far as it agrees with what is observed." He also broke ranks with his Ionian forebears by shifting his focus from how things came to be to how they actually are. By immersing himself in the real-life, flesh-and-blood study of living things, he distanced himself enormously from his predecessors and his teachers. Socrates had said, "I decided to take refuge from the confusion of the senses in argument to determine the truth of reality." Plato condemned the old philosophers to be reborn as birds or fish because they "paid attention to the things in the heavens but in their simplicity supposed that the surest evidence in these matters is that of the eye."

Aristotle even rose above his own work. In the field of logic, he was the master classifier. He invented "the law of the excluded middle"—specifying that a single category cannot embody a given quality and its opposite. This works wonderfully well with numbers—they're either even or odd, prime or compound—and in formal logic. But it runs into trouble in the real world. Chairs blur into couches, orange drifts into yellow; there are birds that don't fly and mammals that do. Aristotle wanted to classify all living things—he saw that as a first step toward making biology a true science. But he realized that he could not sort animals out logically, as if they were coins minted from a set of predetermined molds. Instead, over the course of many years, he compiled similarities and differences, noted signs of close or distant relationships, and tried to make out nature's own groupings. He saw that groups of species share certain important characteristics—they sport feathers or scales, are warm- or cold-blooded, live only in the water or only on land, reproduce by laying eggs or giving birth. From such widely shared features, Aristotle identified many of the major classes of animals we recognize today. His reliance on natural rather than artificial groupings allowed him to classify many species correctly. He realized, for example, that whales and dolphins are not fishes but mammals, and, with surprising objectivity, that apes and humans are closely related.

Of course, Aristotle could not make progress without some theory. From his studies of physics, he placed great emphasis on the qualities of heat and cold. At times this was helpful, for example leading him to use warm- or cold-bloodedness as a defining feature of animals. But in other areas it led him astray. For example, he thought that the primary function of the lungs was to cool the body. He could not completely free himself from the Platonic idea of perfection, and tried to line up animals from the least to the most perfect. And he came to believe that there was something unique to living things, a soul or life force that animated and gave form to matter. That "soul," he thought, was conveyed from the male to the female through the semen.

As we know, Western science fell into a long sleep with the end of the Roman Empire. When the West finally awoke nearly a millennium later, it rediscovered Aristotle and came to revere him as the "master of those who know." Medieval scholars devoured his ideas, and for the most part accepted them uncritically. In many areas, such as physics, science could only advance by throwing away its Aristotelian training wheels and encountering nature directly. That was not the case in biology. Aristotle had already shown the way, not by providing a prepackaged Platonic system, but by offering himself as the model—the first and one of the best—of a naturalist at work. He created biology as a science, asked profound questions, and showed that those questions could be answered, but only through patient and painstaking dialogue with nature herself.

6

Aristarchus, the Forgotten Copernicus

His hypotheses are that the fixed stars and the Sun remain
unmoved, that the Earth revolves about the Sun in the circum-
ference of a circle, the Sun lying in the middle of the orbit.

—*Archimedes*

The rocky, sun-drenched island of Samos gave us two thinkers of unsurpassed vision—Pythagoras and Aristarchus (*c.* 310 B.C.–*c.* 230 B.C.).
Everyone knows of Pythagoras, in part because he founded a Brotherhood that disseminated his philosophical and mathematical ideas.
Hardly anyone has heard of Aristarchus. He inspired no followers,
founded no school, and his greatest idea, that the Earth moves in a circle
around the Sun, vanished like a precious ring thrown into the sea.

The treatise in which Aristarchus presented his heliocentric theory is lost. We know of it only because it was mentioned by a few later
scholars, most notably Archimedes and Plutarch. Archimedes provides the key points of the Sun-centered cosmological model Aristarchus proposed:

- The Sun and the fixed stars do not move.
- The Earth revolves around the Sun.
- The orbit of the Earth is a circle.

25

- The Sun lies at the center of that circle.
- The fixed stars are immensely distant from the Sun and the Earth.

The Roman historian Plutarch, writing two centuries later, supplies a few more details. He tells us that Aristarchus also believed that the Earth rotates once a day, producing the impression that the heavens spin around it. Clearly then, Aristarchus understood that the Earth is a sphere whose daily rotation creates the apparent rotation of the heavens. This may explain why he is credited with having invented a new kind of astronomical instrument called the *skaphe,* a sundial with a bowl-shaped face. Unlike the flat-faced *gnomons* borrowed from the Babylonians, the *skaphe* accurately tracks the Sun's course across the sky. Plutarch also tells us that Aristarchus taught that the Earth moved around "the Sun's circle," that is, the ecliptic. Most scholars assume that, having made the Earth a planet, Aristarchus also put the other planets into orbit around the Sun.

Aristarchus was aware that his model greatly increased the size of the universe. If the Earth did not move, the stars might lie just beyond the Sun, the Moon, and the planets. But if the Earth moved in a vast circle around the Sun, it would sometimes be closer to certain stars, sometimes farther away. Unless the stars were extremely far away, groups of stars should expand or contract as the Earth moved towards or away from them. Since this did not happen, his moving Earth must be wandering in a truly vast universe.

People resent being told that they are not privileged and special. So we should not be surprised that plucking the Earth out of its prized position at the center of the universe did not make Aristarchus a popular man. Cleanthes, who headed the Stoic school when Aristarchus published his radical new model, circulated a tract, "Against Aristarchus," in which he called on the Greeks to indict Aristarchus for impiety. There is no indication that Aristarchus had to defend himself before an actual tribunal, but he must have been uncomfortably aware that his ideas about the heavens had the ability to stir up trouble here on Earth.

Only one work by Aristarchus has come down to us. Although not as earthshaking as his Sun-centered universe, it shows that he was an original thinker, an excellent mathematician, and a serious observational astronomer. In the work entitled *On the Sizes and Distances of the Sun and Moon,* he makes the first mathematically sound attempt to take the measure of the cosmos. He reasoned that when the Moon is exactly half-full, straight lines joining the Earth, Moon, and Sun form a right triangle. If he could measure the angle formed by the lines from the Sun to the Earth and the Earth to the Moon at that moment, he could calculate the relative distances separating the three bodies. He knew from solar eclipses that the Sun and Moon appear the same size from Earth, so once he knew their relative distances he could calculate their relative sizes. Of course, he did not have trigonometric tables in which he could look up the sines and cosines of the angles he measured, but he did have mathematical techniques for approximating them.

Unfortunately, the quality of his measurements did not do justice to his mathematics. He measured the critical Sun-Earth-Moon angle as 87°, while in fact it's just 8′ shy of 90°. He also used an inflated estimate of the apparent diameter of the Moon. As a result, his calculations placed the Sun about 20 times as far from the Earth as the Moon, while the real ratio is closer to 400. Similarly, he estimated that the Sun is just 6 times larger than the Earth, while it's actually 109 times larger. Even though his results were no more than rough approximations, they convincingly proved that the Sun is far larger than the Earth. Since a larger body is not likely to revolve around a smaller one, this discovery may have brought Aristarchus a step closer to his realization that the Earth must revolve around the Sun.

As for his groundbreaking, prescient, heliocentric theory—it essentially disappeared. Archimedes, a generation after Aristarchus, mentioned it, but only in passing. In his work *The Sand Reckoner*, Archimedes wanted to show that he had a way to count the number of grains of sand in the universe no matter how large it was. For him, Aristarchus's theory was just a convenient prop for his numerical tour de force. About a century after Aristarchus, the astronomer Seleucus argued for the heliocentric idea. And after that, nothing. Nobody took the idea seriously for 1,700 years, until Copernicus breathed life into it again. Generations of scholars have asked why the Greeks let this brilliant and powerful idea slip away.

One explanation is competition. We don't know how thoroughly Aristarchus developed his system. But we do know that the cosmology that won out not only kept the Earth securely at rest in the center of the universe, but was elaborated to the point that it was able to account for the motions and positions of the planets with remarkable precision. Over the next four centuries, Apollonius of Perga, then

Hipparchus, and finally the majestic Ptolemy developed the geocentric model of the cosmos until it could accurately predict the movements of the planets far into the future.

To be sure, the Earth-centered system grew from a simple, archetypal concept of uniform circular motion into a baroque monstrosity that only an obsessive scholar could love. Large circles, called deferents, sported smaller circles—the famous epicycles. When those two movements were not enough to approximate a planet's position, the centers of the large circles were nudged away from the Earth, turning them into eccentrics. And when the planets still ran ahead or lagged behind their idealized positions, an even fancier frill was added—the equant. That was a point distinct from the Earth *and* from the center of the deferent, around which a planet moved with steady angular speed. The system was horrendously complicated, but it worked.

Around 150 A.D., Ptolemy gathered together planetary observations dating back to Babylonian times, brilliantly juggled nearly eighty deferents, epicycles, eccentrics, and equants, and produced the definitive astronomical work of the millennium. The Arab scholars who preserved and studied it after the lights went out in the West called it simply *Almagest*—The Greatest. It reigned as the astronomical Bible until long after the death of Copernicus. Ptolemy canonized the geocentric model of the universe. Its only competitor, the speculative and undeveloped heliocentric model, was jettisoned.

In retrospect, the geocentric model won out because it was built on four seductive assumptions, all of which imply that there is something special about our place in the universe or our sense of how things ought to be—and all of which proved to be wrong:

1. The Earth is at the center of the universe, and it doesn't move.
2. Things on Earth are complicated and messy; the heavens are perfect.
3. The fixed stars are the epitome of perfection—they don't move or change.
4. The planets move, but only in perfect circles and at uniform speeds.

Ptolemy made his self-imposed task perfectly clear, writing, "We believe that the object which the astronomer must strive to achieve is this: to demonstrate that all the phenomena in the sky are produced by uniform and circular motions."

As we'll see when we come to Copernicus, it wasn't the fit be-tween Ptolemy's predictions and the observed positions of the planets that bothered Copernicus, nor the system's complexity. Copernicus was outraged that Ptolemy had snuck in nonuniform motion through the use of eccentrics and equants. It would take until 1609 for the tor-tured genius Kepler to finally escape the tyranny of the circle and at last see the planets gliding like serene skaters around an elliptical rink.

Perhaps the ultimate irony in this drama is that the Greeks, pro-found geometers that they were, understood the ellipse perfectly well. As a matter of fact, it was the astronomer-mathematician Apollonius of Perga who made the first thorough study of the mathematical properties of the ellipse. Apollonius must have at least known about Aristarchus and his heliocentric theory. As he struggled with the Earth-centered model, did he ever draw an ellipse and wonder if Venus, Mars, or even the Earth might trace such a path? If that spark had flashed through his mind, Apollonius might have resurrected Aristarchus and saved science from a detour that would take nearly 2,000 years to complete.

7

Archimedes's Physics

Any solid lighter than a fluid will, if placed in the fluid, be so far immersed that the weight of the solid will be equal to the weight of the fluid displaced.

—*Archimedes*, On Floating Bodies

[B]eing perpetually charmed by his familiar siren, that is, by his geometry, he neglected to eat and drink and took no care of his person; . . . he was often carried by force to the baths, and when there he would trace geometrical figures in the ashes of the fire, and with his finger draws lines upon his body when it was anointed with oil, being in a state of great ecstasy and divinely possessed by his science.

—*Plutarch, describing Archimedes*

Historians say that Archimedes (287 B.C.–212 B.C.) disdained the practical. He possessed one of the most original and powerful mathematical minds of all time, and preferred to share his discoveries with the few colleagues who could understood them. Yet, as with many of the mathematicians and scientists of more recent times, the demands of war forced him to turn his genius to practical ends. It was the grim and dangerous real world that guaranteed his fame, and brought about his death.

Archimedes is usually ranked as one of the greatest mathematicians of all time—alongside one or two other incandescent intellects such as Newton and Gauss. Since he valued the pure world of mathematics far more than the mundane and practical, it's probably appropriate that we know far more about his work than about him. He almost certainly would have preferred it that way.

We do know that he lived and died in Syracuse, a Greek city in the southeast corner of Sicily. He was killed in 212 B.C., when Roman legions under Claudius Marcellus conquered Syracuse after a two-year siege. Archimedes was seventy-five years old then, according to Tzetzes, a historian who wrote some fourteen centuries later. If accurate, that would make 287 B.C. the year of his birth. In his brief, playful work *The Sand Reckoner*, in which he invents a method of counting by powers of 100 million, Archimedes mentions that his father was an astronomer who had found ways to estimate the relative size of the Sun and Moon.

Archimedes absorbed two mathematical traditions. He was intimately familiar with the works of Euclid, whose great compilation of geometrical theorems served as a toolbox from which Archimedes drew elements of his own proofs as needed. And he almost certainly studied in Egypt, where he is said to have invented the famed Archimedean screw, a device for raising water by the rotation of a spiral channel. Eighteen hundred years later, Galileo was so impressed by the device that he described it as "not only marvelous, but miraculous." We can't be sure that Archimedes invented the device, but we do know that he stayed on friendly terms with scholars in Alexandria, then a leading center of learning. He sent copies of much of his work to Conon of Samos, a mathematician and astronomer, and to Eratosthenes of Cyrene, famed in astronomy for an early calculation of the Earth's radius, and in mathematics for his "sieve," a way of finding prime numbers. Both lived and worked in Alexandria. Archimedes's work sprang from a soil already tilled by centuries of discoverers, yet with every step he took, he broke new ground.

In 214 B.C., the Romans attacked Syracuse by land and sea. The city did not fall easily or quickly, in large part due to the war machines Archimedes designed. From the physical principles he had discovered and reduced to mathematical clarity, he was able to devise levers and pulleys to catapult immense stones onto the surrounding legions and ships. Roman historians also describe cranes of his design that sank ships by dropping huge weights onto them, or that lifted ships by their prows and then dropped them into the depths. It's easy to imagine the terror such gargantuan devices inspired in the Roman soldiers. Legends rather than history tell us that Archimedes, engrossed in his diagrams, ignored the orders of a soldier and was killed on the spot.

The famous "Eureka" story is also a legend. According to Vitruvius, a Roman architect writing three centuries later, King Hieron II suspected that the maker of a golden wreath he had commissioned substituted silver for some of the gold the king had provided. Hieron challenged Archimedes to determine the purity of the wreath, which, already consecrated to the gods, could not be damaged. The solution came to Archimedes as he submerged himself in a bath, and, overjoyed, he ran naked through the streets shouting, "Eureka"—"I've found it!"

Since we know that Hieron turned to Archimedes in other matters, it's certainly possible that he posed this problem to Archimedes. And there's no doubt that the discoveries Archimedes had made concerning the physical laws governing bodies in water, in particular, his profound Archimedean Principle that a body immersed in a liquid is buoyed up by a force equal to the weight of the liquid it displaces, gave him the tools to solve it. He knew that equal weights of gold and silver occupied different volumes and would displace unequal amounts of water. So he could find the proportion of silver in the wreath by measuring how much water it displaced compared to equal weights of pure gold and pure silver.

It's Archimedes's reaction that doesn't quite ring true. As mathematician Sherman Stein writes, "I doubt that such an elementary observation would have impressed Archimedes enough to stage a celebration, especially when it is compared to his dazzling discoveries about the surface area and volume of a sphere, the center of gravity, and the stability of floating objects."

But let's not throw out the baby with the bathwater. Archimedes's baby, one of the most striking innovations he brought into the world, was a mathematical understanding of buoyancy—his analysis of what makes bodies of any shape, size, and orientation sink or float, capsize or ride securely on the water. These discoveries flowed logically from his earlier discovery of the laws of the lever, his work on the center of gravity, and the famed Archimedean Principle. He applied geometry with enormous power to determine the center of gravity of complex figures and to predict the stability or instability of floating objects. In keeping with his disdain for the practical, Archimedes didn't bother to apply his findings to the boats and ships of his day. But his work forms the foundation of naval design to the present.

In one way, Archimedes was thoroughly modern. As deeply as he valued pure mathematics, he was willing to use other means to master a problem. This was revealed in 1906, when J. L. Heiberg, a Danish mathematical historian, discovered a parchment containing a lost work of Archimedes called "The Method." The copy, from around the year 900, was found in a convent in Constantinople, today's Istanbul. It had been partly erased and overwritten with a Greek Orthodox ritual, but Archimedes's words and diagrams could still be read. To his colleague Eratosthenes, Archimedes described it as "a special method by means of which you will be able to recognize certain mathematical questions with the aid of mechanics. . . . I presume there will be some among the present as well as future generations who by means of the method will be enabled to find other theorems which have not yet fallen to our share."

What Archimedes did was to combine mathematics with mechanics—the laws he had discovered about leverage and the center of gravity of various shapes. With the help of physical models and approximations, he was able to tackle unsolved problems such as the volume and center of gravity of a parabolic solid, the volume of a sphere, and the center of gravity of a hemisphere. His diagrams, models, and thought experiments served as scaffolding from which he could see the solution. True to his mathematical roots, he always went on to prove his findings rigorously. He was particularly inventive and adept at using the method of limits to "trap" a mathematical relationship between converging values. In this, he laid the groundwork for the invention of the integral calculus by Newton and Leibniz 1,800 years later. Archimedes was unique among the Greek philosopher-scientists in utilizing real-world experimentation and in giving us a glimpse of how he worked.

Of all his discoveries, Archimedes was proudest of his proof of a particularly elegant relationship—that the volume of a sphere is two-thirds the volume of a surrounding cylinder. He wanted this to be on his gravestone. Apparently his wish was granted. Almost three centuries after Archimedes's death, Cicero found his tomb, overgrown with thorns, but still topped with a carving of a sphere within a cylinder.

One member of a distant future generation who was inspired by Archimedes's discoveries was Galileo. Among the ancient thinkers Galileo pored over, Archimedes was the one he "read and studied with infinite astonishment." Archimedes inspired Galileo to replace philosophical speculation with experimentation, and to resume the Archimedean task of translating the book of the material world into the language of mathematics. And that, as we will see, shook the world.

8

Ibn al-Haitham Illuminates Vision

[L]ight issues in all directions opposite any body that is illuminated
with any light. Therefore when the eye is opposite a visible object
and the object is illuminated with light of any sort, light comes to
the surface of the eye from the light of the visible object. . . . Therefore the exiting of rays [from the eye] is superfluous and futile.

—*ibn al-Haitham*

The West owes an enormous debt to Islam. Through the bleak centuries following the fall of Rome, Europe stumbled in darkness.
Most of the hard-won knowledge of the Greeks was lost or forgotten.
For six centuries or more, Western science languished. Yet from the
eighth century on, Arabic-speaking scientists, physicians, and mathematicians translated, preserved, and in many cases added to the
knowledge of the ancient Greeks. When the clouds of ignorance and
superstition began to dissipate at the end of the Middle Ages, Islamic
science was there to light the way. Long-lost treasures of Greek philosophy and science returned to the West in Arabic translations, and
along with them came the contributions of great Muslim thinkers
such as the physician Avicenna, the physicist al-Biruni, and the mathematician al-Khwarizmi. Of these, perhaps the most original and influential was ibn al-Haitham (c. 965–c. 1040), known in the West
as Alhazen.

Like Archimedes 1,250 years before him, and Newton 675 years
later, al-Haitham combined great mathematical prowess, the ability to
build general theories from carefully selected facts, and a talent for

experimentation. And, like Copernicus or Galileo, he also possessed great independence of mind. In his autobiography, al-Haitham wrote that since childhood he had been set on finding the Truth, a determination he attributes to "good fortune, or a divine inspiration, or a kind of madness."

There's no doubt that al-Haitham had a great mind. Unfortunately, his command of all areas of knowledge, perhaps coupled with a bit of braggadocio, seems to have gotten him into serious trouble. Al-Hakim, the imperious and murderous Caliph of Egypt around the turn of the last millennium, heard that al-Haitham had boasted that he knew a way to control the flooding of the Nile. Al-Hakim "invited" the scholar to Egypt, an invitation al-Haitham apparently could not refuse. He left his native city of Basra (now in Iraq) and his post as an administrator to travel to Cairo. His feelings must have been more than mixed when the Caliph received him with great honors. Al-Haitham led an expedition up the Nile to the cataracts near Aswan, but soon realized that his scheme was unworkable. According to two different sources, he was so afraid of al-Hakim's anger that he pretended to be insane. Only after al-Hakim died in 1021 could al-Haitham return to his researches. It's said that he earned his living copying ancient manuscripts.

Among those writings were works on optics by Aristotle, Epicurus, Galen, Euclid, and Ptolemy. None of them agreed on the nature of light and vision. The atomist Epicurus argued that objects constantly release thin films of atoms, like the skin shed by a snake. These *simulacra* carry the shape and color of objects into the eye. Aristotle did not believe in atoms, but instead taught that objects send their "qualities" of shape and color through the air and into the eye. The dominant theory, espoused in different forms by Plato, Euclid, and Ptolemy, made the eye the source of vision. Plato taught that the eye sends out a kind of fire. Euclid and Ptolemy wrote that the eye sends out mathematical rays with which it perceives the world. Galen, a physician, took the eye's role a step further. He argued that the brain sends visual spirits along the optic nerve, out through the eye and through the air. This powerful emanation turns the air between the eye and an object into an extension of the perceiver's soul.

Al-Haitham set out to place the study of light and vision on a solid experimental and theoretical foundation. Using a variety of devices of his own design, he systematically studied the transmission, reflection, and refraction of light. He analyzed his findings mathematically. And he completed this tour-de-force by extracting what was useful from the hodgepodge of ancient theories, adding his own findings, and building the first theory of light and vision that made sense physically, mathematically, and physiologically.

Clearly not in awe of the ancients, al-Haitham threw out all the theories that invoked something emanating or radiating from the eye. He observed that bright light, for example from the sun, caused physical pain in the eye, and asked how the eye could fill the entire heavens with light within an instant of being opened. It was clear to him that the eye was designed to be sensitive to light, that light was something real and capable of causing changes in the eye, and that light traveled from luminous or reflecting objects to the eye. He also rejected the atomists' theories of skinlike images floating through the air as physically nonsensical. How could something send copies of itself to hundreds or thousands of people at once? How could the *simulacrum* of a mountain squeeze through the tiny pupil of the eye?

Based on his experiments, al-Haitham teased out the basic facts about light. He used a *camera obscura*, a darkened room into which he could admit light through a small opening, to trace beams of light. He demonstrated that light travels in straight lines, and that light rays from different sources can cross without mixing together. He made spherical, cylindrical, and paraboloidal mirrors, and through systematic experimentation and mathematical analysis studied how they reflect light and form images. In a geometrical proof that would awe generations of mathematicians, he solved the general problem of finding the points on a curved reflector that would reflect light between any two points. He discovered spherical aberration, and that a paraboloidal mirror could focus the light from a distant source perfectly. He performed a series of experiments on refraction, which he correctly linked to a slowing down of light passing into a denser medium. He intuited a version of Fermat's principle of least time. He analyzed both reflection and refraction by the brilliant notion of separating the horizontal and vertical components of a light ray's motion. In short, he brilliantly rebuilt the science of optics.

Still, al-Haitham did not discover exactly how the eye produces the image of an object. He understood that light radiates in all directions from every point on an object's surface, so every point inside the pupil of the eye must receive light from every part of the object. His great insight, the key that unlocked the thousand-year-old mystery of vision, was that the eye must create a one-to-one mapping between each point on an object and a point within the eye. An image was not a thing, it was a transformation, a multicolored map.

The exact mechanism al-Haitham proposed was not quite correct. He argued that the eye responds only to light rays that strike its surface and the front surface of the crystalline lens perpendicularly. He used an idealized model of the eye in which the surface of the cornea and the front of the crystalline lens form parallel arcs. Simple geometry then showed that rays striking the surface of the eye head-on would preserve the spatial relationships of points on an object. A ray reflected from a child's head would strike the lens above a ray from her foot; a ray from her right ear would strike the lens just to the right of a ray from her nose. Al-Haitham justified his theory by arguing that light hitting the eye at an angle would be weakened by being refracted, and by speculating that the eye was simply more sensitive to rays entering it head-on. Johannes Kepler, 600 years later, would finally complete the task al-Haitham started by tracing all the rays reaching the eye, perpendicular and refracted, to an inverted image on the retina. But it was al-Haitham who broke with all the ancient authorities to understand an image as a one-to-one mapping, not a floating skin or the product of projected soul-stuff.

Al-Haitham's great work, the *Treasury of Optics*, reached the West in manuscript form sometime in the thirteenth century. One of the first to understand it was Roger Bacon, whose ideas on optics—and on the experimental method—were shaped by it. The work was first printed in 1572, in Basle. Scientists from Bacon through Kepler were influenced by Alhazen, as they referred to him, as were artists such as Lorenzo Ghiberti and Leonardo da Vinci. Al-Haitham showed Western science not just what to think about light and vision, but how to think scientifically. He was the first scientist to combine systematic experimentation, rigorous mathematical analysis, and physically meaningful theorization. If Bacon was the crystalline lens that transmitted the idea of scientific experimentation to the West, the luminous figure he focused on was that of Abu Ali Hasan ibn al-Haitham.

9

Copernicus Moves the Earth

Copernicus is perhaps the most colourless figure among those who,
by merit or circumstance, shaped mankind's destiny. On the luminous
sky of the Renaissance, he appears as one of those dark stars whose
existence is only revealed by their powerful radiations.

—*Arthur Koestler*

It's difficult to imagine a less likely revolutionary than Nicolaus
Copernicus (1473–1543). He lacked the vision of Pythagoras, the
brilliance of Archimedes, and the fire of Galileo. He was fearful, stubborn, miserly, obsequious to authorities, arrogant toward others, and a
world-class procrastinator. He sampled the ferment of the Renaissance
as a student in Italy, but then retreated to his medieval churchtower
for the remainder of his life. He made few astronomical observations,
preferring to rely on ancient—and unreliable—sources. At a time of
explosive change, he clung to the past in most things. Yet his one great
idea—that the Earth is simply one of the planets circling the Sun—set
in motion a revolution in mankind's perception of the universe and
our place in it.

Copernicus was born in the fortified town of Torun, on the Vistula
River. It was a trading center, and his father, Nicolaus, was one of the
leading merchants. His mother, Barbara Watzenrode, also came from a
well-off family. Her brother Lucas would become bishop and ruler of
Ermland, then an autonomous region between the Kingdom of Poland
and regions ruled by the Teutonic Knights. Lucas played a crucial role

Nicolaus Copernicus.

in Copernicus's life. When Nicolaus was ten or eleven, his father died, and Lucas took responsibility for him, along with his older brother and two sisters. Lucas was an austere and demanding authority figure, but he made sure that the boys were well educated and the girls well married. Nicolaus, serious and studious even as a child, became his uncle's favorite. He enrolled at the University of Krakow in 1491, and later studied in Bologna and Padua. By the time he received his doctorate in Canon Law at the age of thirty, he had studied a bit of everything— Latin and Greek, medicine and law, mathematics and astronomy. In Bologna, he lived in the same house as the University's leading astronomer, Domenico Maria de Novara, and assisted him with his work. After his studies, Copernicus became an administrator at the Frauenberg Cathedral, a lifelong position Lucas arranged for him.

Although Copernicus did not publish his great work, *De Revolutionibus Orbium Coelestium* (The Revolution of the Heavenly Spheres) until his death, the idea came to him in his youth. As he studied the Ptolemaic system, in which the Hellenized Egyptian astronomer Ptolemy had brilliantly systematized five centuries of Greek astronomy, Copernicus found it wanting. The large number of circles and epicycles that it needed to capture the movements of the planets bothered him, but he was even more put off by a basic inconsistency. Ptolemy had canonized the assumption that each planet moved with constant speed at a fixed distance from its center of motion. But in order to reproduce the movements of the planets across the sky accurately, Ptolemy had allowed the speed of a planet to change. In Ptolemy's system, Copernicus complained, "a planet never moves with uniform velocity either in its deferent sphere or with respect to its proper center." He thought there must be a simpler, more consistent way to model the movements of the planets. Copernicus carried that nagging puzzle back to Poland with him, struggled with it obsessively, and found he could solve it through a simple but revolutionary shift— by placing the Sun, not the Earth, at the center of the cosmos.

In concept, Copernicus's answer to Ptolemy was as clear as the heavenly spheres themselves. With the brashness of youth, and in true Renaissance fashion, Copernicus overturned two thousand years of astronomical thinking, along with the primordial perception that Earth stands still while the heavens wheel around it. Not only did he topple Earth from her throne at the center of creation, he sent her spinning through space. Copernicus knew that a few of the Greek philosophers had harbored similar thoughts. The Pythagoreans taught that the Earth, planets, and stars all circle the Central Fire. Heraclides of Pontus had speculated that the Earth rotates like a wheel once a day, and Aristarchus had taught that the Earth revolves around the Sun. But the two greatest astronomers of antiquity, Hipparchos and Ptolemy, had rejected their ideas. Since Ptolemy's day, astronomers and common people alike believed that the Earth stood still while the heavens revolved around it. Copernicus was the first to set Earth firmly among the planets, and the planets in proper sequence around the sun. Our concept of the solar system was born in his mind.

Still, the crystalline ideas of youth often founder on the realities of the world. Copernicus sketched out his "new and marvelous hypotheses" in a handwritten manuscript that he sent to a few of his peers. The year was 1514; he was thirty-six years old. In his *Commentariolus,* or Brief Outline, Copernicus presented his bold and clarifying ideas:

- The Earth is not the center of the universe, although it is the center of the Moon's orbit and of its own gravity.
- The Sun is the center of the planetary system and the sphere of stars. Earth is just one of the planets.
- Since the Moon rotates around the Earth, the heavenly bodies do not share the same center.
- The Earth's distance from the Sun is negligible compared to the distance to the fixed stars.
- The stars, therefore, are vast objects lying at great distances from the Sun and the Earth.
- Although the heavens appear to rotate around the Earth once a day, it is the Earth that rotates on its own axis.
- The Sun appears to move completely around the sky once a year, but this due to the revolution of the Earth around the Sun.
- The complex movements of the planets, which include points of apparent rest and periods in which they seem to reverse their normal march across the sky, along with their brightening and dimming, can be explained by the relative motions of each planet and the Earth.
- Placing the Sun at the center resolves all ambiguities about the order of the planets. The farther they orbit from the Sun, the longer they take to circle it. Mercury is the closest and orbits in the least time, eighty-eight days, while distant Saturn takes thirty years to circle the Sun.

It was a brilliant new synthesis, but the details bedeviled him. Copernicus was determined to use his new model to calculate planetary positions more accurately than Ptolemy had. But now his innate conservatism blocked him. Like Ptolemy, he could only conceive of planets moving on circular paths. Like Ptolemy, Copernicus placed the planets on circles attached to other circles. By the time he'd finished nudging and adjusting, Copernicus had encumbered his crystalline vision with no less than forty-eight cycles and epicycles. "Instead of the harmonious simplicity which the opening chapter of the *Revolutions* promised," writes one biographer, "the system had turned into a confused nightmare."

Maybe it was that nightmarish way his beautiful child had grown up that led Copernicus to hide it away. Or perhaps it was fear. In the preface to *De Revolutionibus*, addressed to Pope Paul III, Copernicus wrote, "Those who know that the consensus of many centuries has sanctioned the conception that the Earth remains at rest in the middle of the heavens as its center would, I reflected, regard it as an insane pronouncement if I made the opposite assertion that the Earth moves. . . . When I weighed these considerations, the scorn which I had reason to fear . . . almost induced me to abandon completely the work which I had undertaken."

In any case, Copernicus completed *De Revolutionibus* by 1530, and then sat on it. Surprisingly, the Catholic Church, through Nicolaus Schoenberg, cardinal of Capua, strongly urged him to publish his new theory. His lifelong friend, Tiedemann Giese, who became bishop of Kulm, pleaded with him for years to publish it. Copernicus finally relented in response to a passionately devoted disciple, Georg Joachim Rheticus. The youthful Rheticus burned with the zeal Copernicus lacked. He sought Copernicus out, studied and mastered his manuscript, and, with Copernicus's blessing, published a detailed summary of it in 1540 called *Narratio Prima*. After more years of cajoling, Rheticus finally convinced Copernicus to allow the full work to see the light of day. Unfortunately, Rheticus was not able to shepherd the book through publication. He entrusted it to a Lutheran priest, Andreas Osiander. When the book appeared in 1543, Osiander, for his own reasons, had added an anonymous introduction couching the book's ideas as mere hypotheses, not to be taken seriously except as a calculating device. Rheticus and Giese were outraged, but the introduction remained, and was accepted as Copernicus's own for many decades.

Despite Osiander's caveats, *De Revolutionibus* was destined to set off a slow-motion revolution. Galileo, Kepler, and Newton built on its foundation, and their work cumulatively destroyed the ancient view of a finite, womblike cosmos with Earth, and the human race, at its center. Cautious old Copernicus managed not to be there to deal with the effects of his ideas.

Kepler, for one, was overwhelmed by the beauty of Copernicus's idea. "I have confessed to the truth of the Copernican view," he wrote, "and contemplate its harmonies with incredible ravishment." But it is legitimate to ask if Copernicus derived any joy from his new vision of the universe. From what we know of his character, it's impossible to imagine him shouting *"Eureka!"* running to tell a friend about his new discovery, or trembling with awe at what he was given to see. But he did appreciate order. That's what seems to have driven him to remake the wheels of creation, and perhaps it also gave him a measure of satisfaction—just visible between the lines of his introduction to *De Revolutionibus*:

> Having thus assumed the motions which I ascribe to the Earth later on in the volume; by long and intense study I finally found that if the motions of the other planets are correlated with the orbiting of the Earth, and are computed for the revolution of each planet, not only do their phenomena follow therefrom but also the order and size of all the planets and spheres, and heaven itself is so linked together that in no portion of it can anything be shifted without disrupting the remaining parts of the universe as a whole.

His friend Giese wrote that a copy of *De Revolutionibus* was given to Copernicus on May 24, 1543, as he lay dying. By then, he was too weak to read. So we will never know what he thought of the mealy-mouthed preface by Osiander. It is clear that Copernicus truly believed that the Earth moves. But it's equally clear that he was deathly afraid of entering the fray of life—of being splashed with the mud of ignorance. Wouldn't it make a lovely story if every hero was truly heroic, every revolutionary a firebrand, every discoverer noble and wise? But it would be just that—a story. Copernicus may have been a dark soul, but he saw the light at the center of the cosmos and made it out to be the Sun. In his own crabbed, cautious, edgy way, he did what Archimedes boasted but could not do—he moved the Earth.

10

Galileo Discovers the Skies

And new Philosophy calls all in doubt,
The Element of fire is quite put out;
The Sun is lost, and th'Earth, and no man's wit
Can well direct him where to look for it.

—*John Donne*, The First Anniversarie *(1611)*

The celestial upheaval Donne described was largely the work of one man—Galileo Galilei (1564–1642). In 1609 he broke through the boundaries of what was known and believed by fashioning a simple telescope and turning it to the skies. In a flood of discoveries, he was the first to observe sunspots and the rotation of the Sun, and mountains and depressions on the Moon, the first to resolve the mysterious Milky Way into separate stars, and, most dramatically, the first to see Jupiter as a miniature planetary system and chart the orbits of its moons. Galileo published his results for all to see in the spring of 1610 in his poetically named *Sidereus Nuncius*—Starry Messenger. In this brief work he turned Copernicus's heliocentric theory into demonstrable fact.

Although a few of the ancient Greek philosophers had speculated that the Earth circles the Sun, Aristotle put his philosophical blessing on the commonsense and comforting idea that the Earth is at rest while the heavens circle around it, perfect and unchanging. Ptolemy, the most influential astronomer of late antiquity, cemented Aristotle's ideas, including the centrally placed Earth, in his immensely influential astronomical bible, the *Almagest*. And there the matter rested until the Polish astronomer Nicolaus Copernicus challenged it in his great book, *De Revolutionibus*, whose publication he delayed until 1543, just before he died.

Galileo knew and understood Copernicus's new system, although he continued to teach Ptolemy's Earth-centered cosmology. In 1597, he wrote to the astronomer Kepler admitting that he supported Copernicus, but still did not dare advocate his ideas publicly. Galileo did not openly state his support of the Copernican system until thirteen years later, in the closing pages of his *Starry Messenger*.

We can get some measure of the forces that kept Galileo quiet by noting that just ten years earlier, the Dominican friar Giordano Bruno had been burned at the stake in Rome, in part for advocating the heretical idea that the Earth circled the Sun rather than resting at the center of the cosmos. Both the Catholic Church and the burgeoning Protestant denominations believed, from their interpretations of the Bible, that the Earth does not move.

Despite his caution, Galileo was well prepared for his revolutionary task. He had shown himself to be one of the leading mathematicians of his day, establishing his reputation at the age of twenty-one when he found a way to improve on Archimedes's methods for finding the center of gravity of irregularly shaped solids. Unlike the Greek philosophers and most of his fellow scholars, Galileo set out to prove or disprove competing theories not just through logic but through experimentation. He observed heavy bodies as they fell (he probably did not drop weights from the Leaning Tower of Pisa, but he painstakingly timed balls rolling down inclined planes). He also designed instruments such as a thermometer and a mechanical calculating device, his geometric and military compass. He was the first to understand that the distance traveled by a falling body increases with the square of time, and that projectiles follow parabolic curves. He made a pre-telescope foray into astronomy in 1604, when he proved that a nova, or new star, in the constellation Serpentarius lay in the supposedly unchanging regions beyond the Moon. Galileo's revolutionary thinking may have been inspired by his father, Vincenzio, who had fought for over a decade to publish his research on a new way of tuning musical instruments, an experimentally based idea that also challenged Aristotle.

With the zeal of a bloodhound hot on a trail, Galileo pushed on with his telescopic observations. By the fall of 1610 he had made close to 100 telescopes. He sent the best of them to influential people throughout Europe, hoping to gain their support. Before the end of the year he had discovered sunspots and from them the rotation of the Sun. This struck another blow against Aristotle's doctrine that celestial objects, unlike earthly ones, were unchanging. Galileo went on to describe the phases of Venus, in the process demolishing one of the strongest arguments against the Copernican system. He was also the first to see and sketch the rings of Saturn, although with his relatively crude telescopes he could not tell if they orbited Saturn or were somehow attached to it.

Rocked by Galileo's stream of discoveries, it's no wonder that by 1611 some of the more open-minded opponents of Copernicus, such as Christopher Clavius, were admitting that the old order, the conceptual foundation of the cosmos for 2,000 years, could no longer serve. Others, however, remained hostile. Galileo demonstrated his telescope to astronomers in Bologna in April 1610. They managed not to see Jupiter's moons. Some philosophers such as Cremonini, an expert on Aristotle, simply refused to look through Galileo's instrument.

In addition to his astronomical discoveries, Galileo also overturned many of Aristotle's views on motion, which had shaped people's understanding of the physical world for nearly two millennia. Aristotle had asserted that heavy weights fall faster than light ones, and subsequent philosophers believed him. Galileo was the first to carry out real-world experiments—dropping and rolling various weights—which contradicted Aristotle, and which founded the scientific study of motion and gravity. But he had already cast doubt on Aristotle's description of falling bodies through an elegant thought experiment: According to Aristotle, a one-pound weight and a ten-pound weight would fall at very different speeds. How fast would they fall, Galileo asked, if you tied them together? According to Aristotle, the lighter weight should hold the heavier one back, so together they would fall more slowly. Tied together, however, they form an eleven-pound weight, which should fall faster. Aristotle's theory contradicted itself, so it could not be right. "What clearer proof do we need of the error of Aristotle's opinion?" Galileo wrote.

Galileo pored over Archimedes's works avidly. Inspired by Archimedes's mathematical treatment of the laws of leverage, the center of gravity of plane and solid objects, and floating bodies, Galileo realized that science must be built on measurement and mathematics. Archimedes had made great strides studying the buoyancy of objects immersed in water. Galileo realized that air, too, is a fluid, and should

Galileo Galilei.

also buoy up objects. He thought this accounted for the small differences he observed in the speed of heavier and lighter objects as they fell. Towards the end of his life he claimed that a chunk of lead and a puff of wool would fall at the same speed in a vacuum. He would have loved the demonstration that David R. Scott, an Apollo 15 astronaut, carried out on the airless surface of the Moon on August 2, 1971. Scott dropped a feather and a hammer at the same time. Television viewers on Earth could see for themselves that the two objects hit the surface together.

As long as competing astronomical theories could be seen as mathematical fictions, necessary for keeping the calendar in order, casting horoscopes, and aiding navigation, the Church could remain above the fray. But Galileo's observations, his careful drawings of Jupiter's retinue of moons, the phases of Venus, and spots defacing the Sun, made the issue real. And to make the situation worse, Galileo insisted on writing in Italian rather than Latin, broadcasting his radical ideas to the masses. It was no longer just mathematics. Either the Earth moved or it did not. Either the Earth, and humanity, lay at the center of creation as the Church decreed, or they wandered insignificantly through space.

Although the Church under Pope Paul V at first tolerated Galileo and his new findings, its tolerance soon vanished. The Inquisition at Rome made the idea that the Earth moved an official heresy in February 1616, and banned Copernicus's book, *De Revolutionibus*, until it was "corrected." The ban did not mention Galileo directly, but the pope sent Cardinal Roberto Bellarmino, a noted Jesuit theologian, to warn Galileo in person to stop teaching or defending Copernican ideas. For seven years Galileo avoided discussing the Copernican system as anything but a hypothesis. Still, his enemies—theologians and competing philosophers who felt threatened by his discoveries, his success, and his combative style—continued to whisper about him to the Inquisition.

In 1623, Maffeo Barberini became the new pope, Urban VIII. Galileo knew Barberini well and considered him an ally, so he went to Rome to talk to him, and returned home with the pope's permission to compare the systems of Copernicus and Ptolemy in writing. Galileo seized the opportunity to write his popular and influential *Dialogue Concerning the Two Chief Systems of the World, Ptolemaic and Copernican*, which appeared in 1632. It presented four "days" of argument between three characters: Salviati, the advocate for Copernicus; Sagredo, a man of common sense; and Simplicio, an Aristotelian.

To Galileo's dismay, the pope, his one-time friend and supporter, turned on him. Historians are still arguing why. Urban may have felt particularly vulnerable due to intrigues within the Vatican and the religious wars sweeping across Europe. The Jesuits, who had long mistrusted Galileo, were becoming more and more powerful. And the pope may have been personally affronted because Galileo had put his words, that fallible humans could never understand the mind of God, into the mouth of Simplicio, clearly the Dialogue's loser. Whatever his reasons, the pope ordered an investigation. Within a month his advisors recommended that Galileo be tried for disobeying the earlier papal warning. The Inquisition ordered Galileo to Rome. The proceedings took more than four months. In the end, the dread Inquisitors placed his *Dialogue* on the index of banned books, and found Galileo "vehemently guilty" of suspicion of heresy.

Galileo's unequal contest with the Church ended with him on his knees. On Wednesday, June 22, 1633, he confessed:

> With a sincere heart and unfeigned faith, I abjure, curse, and detest the said errors and heresies. . . . I swear that in future I will never more say or assert, orally or in writing, anything that might give rise to a similar suspicion about me.

Towards the end of the year the Inquisition let Galileo return to his villa near Florence. Not surprisingly, he was physically and emotionally drained. In letters, he described his violent heart palpitations and loss of appetite. Depressed, he yearned to join his beloved daughter Maria Celeste, a nun who had died in the spring of 1634. By January 1638, Galileo, the great visionary, had gone blind. He wrote to a friend, "This heaven, this Earth, this universe, which I have enlarged a hundred, nay, a thousand-fold beyond the limits previously accepted, are now shriveled up for me into a narrow compass occupied by my own person."

Galileo died on January 8, 1642. But his bold and expansive mind, his indefatigable observation and experimentation, his insistence that all physical phenomena could be explained by mechanical causes and that mathematics was the natural language of science, plus his pugnacious courage, blazed an indelible trail for all scientists to come.

11

Kepler Solves the Planetary Puzzle

I felt as if I had been awakened from a sleep.

— *Kepler*, Astronomia Nova

Talk about a dysfunctional family. Johannes Kepler's grandfather had been mayor of Weil, a walled hilltop town in the wine-growing region of Swabia in southwest Germany. But age and declining fortunes only made him more arrogant, short-tempered, and obstinate. Kepler's grandmother was religious, but "restless, clever, and lying, . . . jealous, extreme in her hatreds, violent, a bearer of grudges." Kepler's father, Henrich, was a tavern keeper and mercenary soldier who eventually abandoned his family. Kepler described him as "vicious, inflexible, quarrelsome and doomed to a bad end . . . treated my mother extremely ill, went finally into exile, and died." His mother, Katherin, was raised by an aunt who was burned as a witch. Kepler depicts his mother as "gossiping and quarrelsome, of a bad disposition." She narrowly escaped the same fate as her aunt. Johannes Kepler (1571–1630) was born prematurely, was nearly blinded by smallpox at age four, and was sick much of his life.

His inner vision, however, was unclouded. It led him to a vista no one had ever glimpsed before—of the true nature of the planetary orbits. For two thousand years, from Pythagoras to Copernicus, the planets had been locked in spherical cages—a projection onto the heavens of a human image of perfection. System after system had been cobbled together to make observations fit those circles within circles.

Kepler smashed the heavenly spheres and set the planets on their true elliptical paths, orbiting in harmony with nature's laws, not man's.

Despite Kepler's chaotic childhood, his intelligence shone through. In 1859, he was awarded a scholarship to attend the Lutheran seminary at the University of Tübingen. He had strong religious feelings and hoped to become a theologian. His schoolmates, predictably, taunted, ostracized, and beat him. His unpopularity continued throughout his school years. We know the details of his misery because he described himself as pitilessly as he wrote about others. He saw himself as expressing a welter of contradictory tendencies—greedy, yet content with the simplest food, always seeking approval, yet alienating others with his malice, religious to the point of superstition, yet, in his own eye, morally flawed. Still, it was at the university that he met and formed a lifelong relationship with the astronomy professor Michael Maestlin, who introduced him to the heliocentric theory of Copernicus. We sense Kepler's passionate nature in his reaction to Copernicus. "I have confessed to the truth of the Copernican view," he wrote, "and contemplate its harmonies with incredible ravishment."

There was something in Kepler that would not let his life be destroyed by illness, inner demons, nor the chaotic times in which he lived. In part it was his intellect. He could not help throwing himself enthusiastically into the depths of almost any subject, especially astronomy and mathematics. In part it was his mysticism—he shared with the Pythagoreans the intuition that elegant mathematical regularities were the foundations of the cosmos. Like Pythagoras, he was driven—both would say divinely inspired—to seek out the harmonies underlying the chaos of experience. And that's what Kepler eventually accomplished, although it took him many years, led him down many wrong paths, and required an immense amount of effort.

In the pages of his 1596 book, *Mysterium Cosmographicum*, or *Cosmographic Mystery*, appears some of the most original thinking about the planets since antiquity. The book presents an idea that gripped his mind with the force of delusion for many years—that a nest of the five regular solids explains the number and spacing of the six planets. The idea was seductively beautiful, but wrong, as he himself later realized. The book's greatness lies in the questions Kepler asked and in the physical, rather than metaphysical intuition with which he answered them. He wondered why the planets orbit at their particular distances and speeds. His answer, at first couched in mystical terms, was that some kind of force must emanate from the Sun to sweep the planets along, a force which diminishes with distance much like the light from the Sun. His youthful genius had propelled him not simply to describe the movements of heavenly bodies, but to grope toward a physical cause of the phenomena. It was something nobody had attempted since antiquity.

The *Mysterium* brought him to the attention of Tycho Brahe, the Danish astronomer who was painstakingly making the most accurate naked-eye observations of the stars and planets ever accomplished. Kepler became Tycho's assistant, and, when Tycho died on October 24, 1601, took on his role as imperial mathematician at the court of Rudolph II. But more importantly, Kepler gained access to Tycho's hoarded treasure—his years of observations of the planets, and especially of Mars. "I confess that when Tycho died," Kepler wrote, "I quickly took advantage of the absence, or lack of circumspection, of the heirs, by taking the observations under my care, or perhaps usurping them."

Kepler focused on Mars because its orbit was the most difficult to explain using any of the existing theories. Kepler knew that Tycho's observations were accurate to within 2' (two minutes of arc, or two-sixtieths of a degree). But calculations based on circular orbits inevitably produced errors of up to 8'. By the end of 1604, after many wrong turns, Kepler realized that Mars could not trace a circular orbit, and in fact moved in some kind of oval. His faith in Tycho's observations inspired him to overturn a view of the heavens that had hypnotized mankind for 2,000 years. "Because these 8' could not be ignored," Kepler wrote, "they alone have led to a total reformation of astronomy."

Kepler published his findings in 1609, in his aptly named *Astronomia Nova*, or *The New Astronomy*. There he presented his first two laws of planetary motion. He added significantly to his work in his 1618 book, *Epitome of Copernican Astronomy*. By then he had discovered his third law. These profound insights are now known as Kepler's laws, or simply as the laws of planetary motion:

1. A planet orbits in an ellipse, with the Sun at one focus of the ellipse.
2. A line drawn from the Sun to the planet sweeps out equal areas in equal times.
3. The square of the time it takes a planet to orbit the Sun is proportional to the cube of the planet's average distance from the Sun.

The first law put the planets on their true paths. Like Alexander the Great severing the Gordian knot, Kepler cut through a two-thousand-year-old tangle of theories and observations with a single, elegant stroke.

The second law pleased Kepler immensely. In an unexpected and beautiful way, it transformed the ancient belief that a planet orbits at a uniform speed. Not so, Kepler found. When a planet is farther from the Sun, it moves more slowly, and when it is closer to the Sun, more quickly. However, it speeds up and slows down in perfect proportion to its distance from the Sun.

The third law, which took him two decades to discover, was equally momentous. For the first time, it gave astronomers a precise way to determine the relative sizes of planetary orbits. Kepler was awestruck at this newfound regularity in the heavens. Of it he wrote, "Has not God Himself waited six thousand years for someone to contemplate His work with understanding?"

Kepler's accomplishments were so profound, and arrived at through such a unique and personal process, that scholars count him as one of the few scientists whose work could not have been accomplished by someone else. "It can be said of Kepler, as of very few great scientists," writes historian of astronomy Bruce Stephenson, "that what he accomplished would never have been done had he himself not done it." Charles Gillispie is equally awestruck. "His was one of the great elastic feats of the human mind, comparable in its novelty only to the enunciation of relativity, which also altered the fundamental shape science finds in nature."

Kepler single-handedly invented physical astronomy; he was the first scientist to place astronomy on the foundation of physics. Remarkably, Kepler made his great discoveries based on physical theories that were fundamentally wrong. "This physics was flawed at the foundations, for he lacked the modern concept of inertia," says Stephenson. Kepler did grope toward an improved understanding of gravity. He reached the point of seeing it as the mutual attraction of like bodies, and even understood the link between the Moon's gravity and the tides. However, he did not glimpse gravity's basic role in the movement of the planets—that had to wait for Newton, a half-century later. It's a tribute to Kepler's intellectual honesty, and to his immense respect for Tycho's observations, that he could set aside his mystical notions of the structure of the solar system *and* push past his erroneous ideas of what moved the planets to arrive at the truth. "The physical principles themselves . . . were wrong in virtually all of their

Johannes Kepler.

particulars," writes Stephenson. "The astronomy was startlingly good, far the best that had ever been created."

Intriguingly, Kepler never lost his deeply mystical world view. In 1619, he published *Harmonice Mundi* (Harmonies of the World), in which he reawakened the ghost of Pythagoras, attempting to link geometry, music, astronomy, and astrology to underlying geometrical ratios—the harmonies of the world. Although Kepler was skeptical of many of the claims of astrology, he practiced it—quite successfully—throughout his life. There's no doubt that Kepler's laws helped guide Newton toward his new and correct understanding of gravity and inertia, but it's equally clear that Kepler would not have been at home in the clock-work universe that Newton's physics created.

Kepler's personal life continued to be anything but harmonious. He lived in an era of religious upheavals and had to move several times because of religious issues. At one point the Lutherans denied him communion; later the Catholics confiscated his library and forced his children to attend mass. His first wife died, as did six of his children. From 1615 through 1621 he had to devote much of his time and resources to helping his mother fight charges of witchcraft. He succeeded in getting her exonerated, but she died a few months later. Kepler followed her in 1630, at the age of fifty-eight. Even his grave was lost in the chaos of the Thirty Years' War.

Kepler was many things: brilliant, driven, long-suffering, exquisitely self aware. We owe him a profound debt. Inspired by his unique genius, he pushed through darkness and chaos all his life to reveal to us the elegant harmonies to which the planets dance.

12

Van Leeuwenhoek Explores the Microcosm

In the year 1675, about half-way through September, I discovered living creatures in rain, which had stood but a few days in a new tub that was painted blue within. This observation provoked me to investigate this water more narrowly; and especially because these little animals were to my eye, more than ten thousand times smaller than the animalcule . . . which you can see alive and moving in water with the bare eye.

—*Antony van Leeuwenhoek, October 9, 1676*

Like Galileo, Antony van Leeuwenhoek (1632–1723) was the first to explore a new world. Not the vast cosmos that can be seen through the telescope, but the even stranger world of the invisibly small. The list of Leeuwenhoek's discoveries is overwhelming. He was the first to see and describe one-celled plants and animals, bacteria, blood cells, and spermatozoa. He was fascinated with the mystery of reproduction, and traced the life cycles of many of the microbes he discovered, as well as studying the seeds of many plants and the reproductive anatomy of such familiar animals as frogs, dogs, and fleas. Leeuwenhoek was the first to measure the microscopic world, accurately estimating the size, volume, and population of the inhabitants of the water-drop worlds he studied. He was also the first to see blood pulsing through capillaries in rhythm with the beating heart. In the course of

Antony van Leeuwenhoek.

---

the nearly fifty years he spent peering through the hundreds of microscopes he made, he opened up whole new fields of study—of cells,
blood, and the tissues of plants and animals, to name a few.

Leeuwenhoek's discoveries would amount to a towering accomplishment for any scientist. What makes them even more amazing is
that he was an amateur, completely self-taught. His very ordinary education prepared him adequately for his career selling silk, buttons, and
ribbons to the citizens of his home town, Delft, in Holland. He taught
himself enough mathematics to moonlight as a surveyor and the town
wine measurer. Leeuwenhoek did not see much of the world; he seems

to have ventured to Antwerp, and once as far as England. He never learned to speak, read, or write any language but Dutch, so almost all the scientific literature of his time was inaccessible to him. Anyone else would have been lost from the start. But Leeuwenhoek's drive to understand nature, the quality of the instruments he made, the keenness of his vision and observations, and his devotion to the truth kept him on track. Microbiologist Clifford Dobell writes, "I have unbounded admiration for Leeuwenhoek because he heard and interpreted things that I, unaided, could never have discovered, and hit on problems . . . of which neither he nor I can ever know the final solution."

There's little in Leeuwenhoek's early life to foreshadow his later discoveries. He was the fifth child born to a family of tradespeople in the tidy, busy town of Delft. His father died when Antony was five. Three years later, his mother remarried and sent him to live with one of her brothers, a sheriff in a nearby town. At sixteen, Antony was sent to Amsterdam where he quickly mastered the draper's trade. He returned to Delft in 1654, when he was twenty-two. There he opened a business, married Barbara de Mey, and bought a house. Barbara bore five children in the dozen years before her death. Only one of them, Maria, survived childhood. Like Galileo's daughter Maria Celeste, Leeuwenhoek's Maria remained devoted to her father throughout his life. He remarried in 1671. Cornelia, his second wife, died in 1695. You can get a glimpse of Leeuwenhoek's world in the paintings of his contemporary, Jan Vermeer. Luckily, it was a time in Holland when social class was not a barrier to success.

Leeuwenhoek was nearing forty before he built his first microscope. Ever practical, he had reined in his passion for nature until he achieved a measure of financial security. He did not invent the microscope—a few instruments had been around since before he was born. But those that were available frustrated anyone trying to use them with their low-resolution, highly distorted images. Self-taught and stubbornly independent, Leeuwenhoek didn't bother with the two-lens compound microscopes others were using. His instruments were simple and precise, like Leeuwenhoek himself. For each new microscope, he ground and polished a single small lens of extremely short focal length that he sandwiched between two thin metal plates, each perforated with a small hole. He mounted the object he wanted to study on a pin, and used adjusting screws to maneuver it into focus.

Leeuwenhoek made and chose his lenses so carefully that observers who looked through his instruments during his life, and even years after his death, inevitably lauded the brightness and clarity of the images they produced. Of his few surviving instruments, the most powerful magnifies 270 times, and can resolve features separated by 1.4 microns (a micron is one-millionth of a meter). Leeuwenhoek never showed anyone his best microscopes, and hinted about a secret method that allowed him to make out even smaller objects. Dobell suspects that Leeuwenhoek discovered dark-field illumination, a powerful technique in use today, which makes objects glow against a black background.

Like any explorer in virgin territory, Leeuwenhoek followed where his curiosity led him. He studied the mouth parts of bees and plaque from his own teeth; hair from his beard and the hairs of fleas; red blood cells and the cells of plants; salt and pepper; sperm and eggs; capillaries and nerve fibers; the lens of the eye and the structure of teeth; and, most famously, microbes. The "animalcules" that danced and whirled and darted before his eyes had remained invisible and for the most part unsuspected since the dawn of time. Leeuwenhoek saw them for the first time, described them, drew them, counted them, measured them, and told the world about them.

Leeuwenhoek could easily—and happily—have worked in obscurity and shared his discoveries with a few curious townspeople. But a Delft physician, Reinier de Graaf, alerted the Royal Society of London to his work. The Royal Society, then just a decade old, did not demand scholarly credentials of its scientific correspondents. As its charter said, "to have sound Senses and Truth is with them a sufficient Qualification." For the next fifty years, Leeuwenhoek's exploration of the microscopic world appeared in their *Transactions* in the form of letters, written by him in Dutch, illustrated by one or more unidentified collaborators, and translated by the Society.

Leeuwenhoek's letters read far more like an explorer's journal than polished scientific reports. They ring with the delight he felt. For example, in his sixth letter he describes the discovery of what we now know as protozoa, or one-celled animals: "Among these there were, besides, very many little animalcules, whereof some where roundish, while others, a bit bigger, consisted of an oval. On these last I saw two little legs near the head and two little fins at the hindmost end of the body. . . . And the motion of most of these animalcules in the water was so swift, and so various, upwards, downwards, and roundabouts, that it was wonderful to see."

Read one by one, Leeuwenhoek's letters can seem to be little more than vivid but rambling descriptions, a jumble of unrelated observations. However, he revisited the same subjects again and again through his long scientific life, each time improving his instruments, refining his observations, and testing his ideas. Taken as a whole, the letters form the first, and for many decades, the best map of the microscopic world.

Leeuwenhoek had little respect for most of the theories of his day. He found that they failed to match what he saw with his own eyes. The keenness of his observations was matched by the finesse of his technique. For example, he was able to dissect and study the testes of a flea. Awed by the perfection of form and function revealed in his "little animals," he scorned the theory, accepted since Aristotle, that so-called lower organisms could spontaneously form from non-living material—for example, that a mixture of wheat and rags could spawn mice. "A flea or a louse can no more come into being from a little bit of dirt than a horse from a dunghill," he wrote.

It took decades, and in some cases centuries, for science to incorporate and build on his discoveries. He noticed "an incredible number of living animalcules" in the "unclean matter" on the decaying teeth of an old man, and spirochetes in his own stools during a bout of diarrhea. A century and a half passed before Pasteur and Koch would pin down the role of bacteria in disease. Leeuwenhoek found that vinegar instantly killed most of his animalcules. But physicians did not start saving lives through antisepsis until the nineteenth century. He observed sperm in thirty kinds of animals, and knew that sperm had to penetrate an egg to fertilize it. Yet, in 1814, a leading scientist was still arguing that spermatozoa were parasites that had nothing to do with fertilization.

While the studiously humble words of some great scientists deserve to be taken with a grain of salt, Leeuwenhoek's whole life illuminates his description of what he did:

> As I aim at nothing but Truth, and so far as in me lieth, to point out Mistakes that may have crept into certain matters . . . and if (others) would expose any Errors in my own Discoveries, I'd esteem it a Service; all the more, because 'twould thereby give me Encouragement towards the Attaining of a nicer Accuracy.

13

Newton:
Gravity and Light

And from these principles well understood, it will now be easy to determine
the motions of the celestial bodies among themselves.

—*Isaac Newton*

Nature was to him an open book, whose letters he could read without effort. . . .
He stands before us strong, certain, and alone.

—*Albert Einstein*

Einstein was not exaggerating—Newton does stand alone. Isaac Newton (1642–1727) was the source of four revolutionary scientific advances—the calculus in mathematics, the understanding of light and color in optics, his three laws of motion, and the theory of universal gravitation. Not only did he give mankind incomparable intellectual tools, but he demonstrated for everyone to see that the universe is governed by universal laws that the human mind can find and use. Martin Rees, one of today's leading astrophysicists, says simply, "Newton was the pre-eminent scientific intellect of the last millennium."

There's nothing in Newton's background that explains or predicts an intellect that would grow to encompass the workings of the universe. His father, described as "a wild, extravagant, weak man," died before Isaac was born. Isaac was born prematurely, and his childhood was marred by an early separation from his mother. His mother remarried before he turned two, and handed him off to her parents. Although she lived just a few miles away, she remained out of reach to the "sober, silent, thinking" boy until her second husband's death nine years later. Some biographers believe that the separation accounts for Newton's lifelong emotional problems—tantrums as a child, rage at his mother and stepfather, a paranoid overreaction to criticism, two full-fledged breakdowns, and an implacable, anger-driven vindictiveness that makes Victor Hugo's Javert seem softhearted.

Luckily, a few people saw something in this sullen young man— among them John Stokes, his schoolmaster, and William Ayscough, his uncle. They convinced his mother to give up her doomed effort to make a farmer of him and send him back to school. Ayscough had graduated from Trinity College, Cambridge, and saw to it that New-ton was admitted there. At eighteen, Newton was two years older and far more serious than his classmates.

On the surface, Newton made adequate progress, absorbing the classical authors and ideas then being taught. Although Copernicus, Kepler, Galileo, and Descartes had made groundbreaking discoveries in the previous 150 years, Trinity still taught Aristotle's qualitative physics and Euclid's geometry. His mathematics professor, Isaac Bar-row, initially saw Newton as an indifferent student, noting that he hadn't yet mastered Euclid. What neither Barrow nor anyone else knew was that Newton had already mastered the seminal works of the new sciences and was pushing beyond them into unexplored intellec-tual territory. But the young Newton, as he would all his life, kept his discoveries to himself, locked away in his notebooks. The motto he wrote in his first notebook seems appropriate: "My best friend is truth."

The plague came to England in 1664, and in the fall of 1665 Cam-bridge closed its doors. Newton spent the next two years at home in the hamlet of Woolsthorpe. In that brief retreat, Newton made such pro-found discoveries that scholars call them "the miraculous years." Early on he developed the binomial theorem, a way to calculate an algebraic expression as the sum of an infinite series of terms. A few months later, he devised his "method of tangents," a key step toward the calculus, a new kind of mathematics, which for the first time would allow mathe-maticians to deal with continuously varying quantities—that is, with change. Within six months Newton had developed "the direct method of fluxions," what we now call differential calculus.

He was curious about light and color. In a brilliant series of exper-iments with prisms, he demonstrated for the first time that white light is a mixture of rays of different colors. And then there was gravity. Newton later wrote, "and in the same year I began to think of gravity

SIR ISAAC NEWTON
(Physicien et Philosophe),
Président de la société royale de Londres.
Né à Woolsthorpe dans le Comté de Lincoln le 25 Décembre 1642.
Mort à Londres le 20 Mars 1727.

Isaac Newton.

extending to the orb of the Moon . . . and compared the force requisite to keep the Moon in her orb with the force of gravity at the surface of the Earth."

As brilliant as he may have been, ideas did not simply leap into his mind. Newton later "explained" how he came by his insights. "By always thinking unto them," he said. "I keep the subject constantly before me and wait until the first dawnings open little by little into the full light." One of those dawnings, he told an early biographer, was seeing an apple fall. The story may or may not be true. But it is clear that it was only Newton's vision that could see the fall of an apple and the orbit of the Moon as governed by the same laws, and only his mind that could enlarge that insight into the theory of universal gravitation.

It took Newton twenty years to weave together all the pieces into his masterwork, *The Mathematical Principles of Natural Philosophy*. It appeared in 1687, but only after determined prodding by his friend, the astronomer Edmond Halley, who also paid for the book's publication. In the *Principia*, Newton gathered together the scattered and often contradictory observations and theories about how the world works that had been developed since the Greeks, smelted and refined them in the crucible of his mind, and poured them out as a seamless stream of understanding. He unerringly rejected the slag—the easy confusion of magnetism and gravity, circular inertia, Galileo's theory of the tides, Kepler's rotating spokes of force, and Descartes's swirling vortices. What he kept was pure gold.

In the *Principia*, Newton systematically showed how his three laws of motion plus his law of universal gravitation explained everything that Galileo had discovered about falling bodies, everything Kepler had learned about the motions of planets, not to mention the orbits of the Moon, planets and comets, the behavior of fluids and bodies moving in them, the precession of the equinoxes, and the tides. The book was immediately recognized as a masterpiece, and continues to be seen as the most important scientific book ever written.

The Law of Universal Gravitation is the *Principia*'s centerpiece. Newton first glimpsed the law of gravity when he was twenty-four. In the *Principia*, two decades later, he showed that gravity pervades the universe. Every particle of matter attracts every other particle, he asserted. And he was the first to be able to measure the strength of that attraction: simply multiply the particles' masses and divide by the square of the distance between them.

Single-handedly, Newton had distilled crystal-clear definitions of gravitation, force, mass, momentum, and acceleration, and provided the world with the mathematical tools for dealing with motion and change. He appeared to have written the rule book of the universe.

Newton set his gleaming clockwork universe on what he believed was the firmest of foundations—absolute space and absolute time, allowing for absolute motion. "Absolute, true and mathematical time flows in virtue of its own nature uniformly and without reference to any external object," he wrote, and "absolute space, by virtue of its own nature and without reference to any external object, always remains the same and is immovable." That magnificent edifice stood unscathed for 218 years. It was only in 1905, with the publication of Einstein's theory of special relativity, that its crystalline walls melted into a subtler understanding of nature.

Newton was not only one of the supreme theorists and mathematicians of all time, he was also an inspired experimenter. His *Opticks*, published in 1704, is remarkable not just for its fundamental discoveries, but for the effortless way in which each experiment leads inevitably to the next. Newton proved that white light was actually a mixture of light of different colors, that the degree to which light is bent or refracted when it flows from one medium to another is governed by its color, and that light separated into its different colors can be recombined to produce white light again. Even Einstein was impressed, writing, "The conceptions which he used to reduce the material of experience to order seemed to flow spontaneously from experience itself, from the beautiful experiments which he ranged in order like playthings."

Throughout his life Newton poured as much of his formidable intellectual energy into alchemical researches and theological speculations as he did into the scientific work we now value. Although he brought the same keenness of mind, originality of thought, and tireless effort to these studies as he did to his studies of motion, gravity, and light, they seemed to lead nowhere. Still, some scholars argue that his deep involvement with the mystical Hermetic tradition may have given him the model for universal forces between objects. While his years of alchemical experiments never transformed base metals into gold, he may have transmuted mystical notions of attraction into universal gravitation. No doubt he included these curious investigations in his famous comment about discovery: "I do not know what I may appear to the world, but to myself I seem to have been only like a boy playing on the sea-shore, and diverting myself in now and then finding a smoother pebble or a prettier shell than ordinary, whilst the great ocean of truth lay all undiscovered before me."

In the last decades of his life Newton invested the same energy and intellect into making sure that he—and he alone—would receive full credit for his discoveries. His rancorous feud with Robert Hooke, who was foolhardy enough to challenge Newton as author of the inverse-square law of gravity, ended only in 1703, when Hooke died. But the enemy who incurred the full measure of Newton's wrath was Gottfried

Wilhelm Leibniz, who independently developed the calculus, and, unlike the secretive Newton, published it promptly. It was a battle of titans, with the prize being credit for one of the greatest mathematical advances of all time. Most scholars now believe that Newton developed "fluxions"—his version of the calculus—before Leibniz, but that Leibniz created his system independently. Leibniz did manage to get the last word; it's his user-friendly set of symbols that every calculus student learns. For three decades, their battle for priority raged across Europe, drawing scientists, diplomats, and even royalty into the melee. Newton publicly defamed Leibniz every chance he got. We now know that Newton secretly wrote the supposedly impartial report of the Royal Society on the question of priority, plus the anonymous summary and review of that report, plus scores of critical articles which he coerced younger scientists into publishing as their own. Leibniz died a broken man in 1716. Even that did not cool Newton's vendetta against him.

If there is a measure of mind, it may be its ability to create order from chaos—to see the tiger camouflaged in the forest's jumble of light and dark. Whatever his psychological quirks and character flaws, Newton has to be ranked at the top of that scale. No one person has deciphered more of nature's hidden truths. His ideas, in mathematical form, allowed us to launch the first satellite, step onto the moon, navigate to the planets, and weigh and map the movements of distant planets, stars, and galaxies. Like the universal gravitation he discovered, his mind reached to the edges of the universe, and brought order to everything it touched.

14

A Breath of
Fresh Air

The injury which is continually done to the atmosphere by
the respiration of such a large number of animals . . . is, in
part at least, repaired by the vegetable creation.

—*Joseph Priestley, 1778*

It's not difficult to trace Joseph Priestley's lifelong independence of
thought. Priestley (1733–1804) was born into a family of Calvinist
Dissenters struggling against the Church of England. Even within his
family, Priestley seemed destined to make his own way. The oldest of
six children, he lived first with his grandparents, and after his mother
died went to live with a childless aunt. From childhood on, Priestley
dealt with problems by throwing himself into his studies and his work.
He was an avid student, and supplemented what he learned in school
with systematic studies on his own. To prepare to be a minister he
taught himself Hebrew, Chaldean, and Syriac. When it looked as
though he might become a businessman, he learned French, German, and Italian. His interests were enormously wide. He studied,
wrote about, and made significant contributions to language, history,
theology, education, and several branches of "natural philosophy."

Priestley was approaching the age of forty before he took on the
field of chemistry. What interested him from the start was air. Based
on assumptions made by the ancient Greeks, air was still believed to
be a simple, unalterable substance, one of the four basic elements.
Priestley tackled the subject with his usual energy. He reviewed

much of what was thought or known, repeated key experiments, and then devised experiments and approaches on his own. His studies, like Leeuwenhoek's explorations a century earlier, had a freewheeling quality. Like Leeuwenhoek, Priestley was often surprised and delighted by what he found, and reported his successes and blunders with equal openness. He started with washtubs and glassware borrowed from his wife's kitchen. But he was skillful at designing new apparatus as needed, and—with help from friends such as the ceramicist Josiah Wedgwood—eventually created one of the best-equipped chemical laboratories in the world.

Two great discoveries stand out from the huge body of Priestley's chemical work—oxygen and photosynthesis. He first generated oxygen as part of a systematic study of the "airs" produced by heating a variety of substances. He would collect the resulting air over water or mercury, then analyze it with simple tests: What did it look like? Would it burn? Would it keep a mouse alive, and for how long? Would it be absorbed by water? On August 1, 1774, Priestley heated some red *mercurius calcinatus*—what we would now call mercuric oxide—with a newly acquired twelve-inch magnifying lens. He found that a candle burned brilliantly in this new air, and a smoldering splinter of wood sparkled explosively. In March of the next year, he was surprised to find that a mouse lived twice as long in a container of the new air as it would have in ordinary air. He breathed some of it himself, and noted the light and easy feeling it produced. After these and many other experiments, Priestley convinced himself that he had found the part of the atmosphere that supported combustion and life. With remarkable foresight, he speculated that it might eventually be used in medicine, or even as a luxury product.

We now know that the discovery of oxygen was "in the air." In Sweden, an impoverished apothecary, Carl Wilhelm Scheele, was pursuing his own experiments "on air and fire." Although his results were not published until 1777, Scheele generated and observed the properties of oxygen just before Priestley. Still, Priestley's multiple discoveries and the sheer volume of his experimental work put his findings, rather than Scheele's, into the scientific mainstream.

Priestley called his lively new substance "dephlogisticated air." The term phlogiston had been used since the early 1600s to explain combustion. In his 1723 textbook, *Fundamenta Chymiae*, Georg Ernst Stahl defined phlogiston not as fire itself, but as "the matter and principle of fire." Phlogiston escaped from any substance as it burned.

Flammable substances such as oil and charcoal possessed a lot of it; soot was almost pure phlogiston. When metals burned, they too gave off phlogiston. Phlogiston could be put back into the resulting product, for example, calyx of zinc, by heating it with charcoal, in order to reclaim the original metal. Phlogiston had some quirky properties — most notably that adding it to a substance mysteriously made the material lighter.

We now know that phlogiston theory dealt with combustion "backwards." Gaining phlogiston meant losing oxygen, which of course explains the loss of weight. Losing phlogiston meant adding oxygen. Zinc burns in air, producing zinc oxide (calyx of zinc). Heating zinc oxide over charcoal does not add phlogiston; it releases oxygen. But just as Ptolemy's cycles and epicycles did a remarkably good job of explaining and predicting the movements of the planets, phlogiston was able to explain many chemical reactions, at least in qualitative terms. It even led to many fruitful experiments, including much of Priestley's work. It was not until 1777 that Antoine Lavoisier's careful measurements of the weight gained and lost in similar reactions led him to make oxygen, not phlogiston, the key player in combustion and respiration. Even then, phlogiston clung to life as an explanation of the light and heat given off in chemical reactions. It took until the mid-1800s before a clearer understanding of energy finally pushed phlogiston off the scientific stage.

The discovery of oxygen catalyzed the chemical revolution. It led to an understanding of elements, compounds, and the conservation of matter in chemical reactions. And those concepts in turn led to today's dyes, plastics, fertilizers, and designer drugs. Priestley lived to see, if not recognize, the revolution's birth. It took the vision of Lavoisier (who was soon to die on the guillotine) to provide a new explanation for Priestley's key discoveries in terms of the reactions of Priestley's new element, oxygen. It's ironic that Priestley, whose revolutionary religious and political beliefs nearly lost him his life, clung to phlogiston theory until his death. One of his last, defiant publications was *The Doctrine of Phlogiston established and that of the Composition of Water refuted*. It's equally ironic that both Priestley and Lavoisier appealed to a similar scientific esthetic of simplicity to support their arguments. To Priestley, Lavoisier was conjuring up a Pandora's box of new elements. To Lavoisier, Priestley was clinging to a redundant, physically meaningless substance. History, we now know, took Lavoisier's side.

Priestley also discovered photosynthesis, although he did not fully understand it. It was common knowledge that people and animals

Joseph Priestley.

would die if kept too long in a space without fresh air. Priestley, like others, assumed that plants would react the same way. But he believed in checking assumptions experimentally. In 1771, he grew a sprig of mint in a sealed container over water and was surprised to find that it thrived. He tested the air in the container, and found that it supported a candle flame and kept a mouse alive. Intrigued, Priestley produced "vitiated air," by burning a candle in it, then waited to see if a growing plant would make the air "good" again. He even experimented with different kinds of plants to see which made air breathable again most quickly. He came back to these experiments in 1777 and 1778, and now observed bubbles of what proved to be "very pure air" — oxygen — coming from the stems of plants, from green algae, and inside the bladders of seaweed.

Priestley gradually realized that sunlight was necessary to produce the green matter of plants and algae, which in turn produced pure or dephlogisticated air. Remarkably, his experiments also led him to understand that green plants take carbon dioxide (which he knew as "fixed air") from the atmosphere, and return oxygen to it. However, he was slow to perceive the crucial function of light in the production of oxygen by green plants. He later learned, and graciously conceded, that Jan Ingenhousz made that connection earlier than he did.

Like Galileo, Priestley fell victim to persecution. It was not his scientific discoveries but his political beliefs that put him at odds with both church and state—and nearly cost him his life. Priestley was an outspoken and at times passionate advocate of freedom of thought and speech, and became one of the most visible supporters in England of both the American and the French revolutions. Despite angry denunciations from people in power and from commoners, he pushed on even when someone less determined might have pulled back. He helped organized a public dinner on July 14, 1791, to commemorate the two-year-old French Revolution. This gave his enemies the excuse they needed to strike. That night a drunken "church-and-king" mob, fed incendiary lies and given free rein by local authorities, sacked and burned Priestley's New Meeting church in Birmingham, and then thundered on to do the same to his home, library, and laboratory. Priestley and his family fled to safety just ahead of the mob. He was close enough to see the flames and hear the sounds of his home being destroyed. Priestley was never able to return to Birmingham. A letter he wrote to his one-time neighbors a few days after the mob attack is a study in the restrained passion of a deeply wounded, deeply disciplined man:

> You have destroyed the most truly valuable and useful apparatus of philosophical instruments that perhaps any individual, in this or any other country, was ever possessed of . . . for the advancement of science, for the benefit of my country and mankind. . . . But what I feel far more, you have destroyed manuscripts which have been the result of the laborious study of many years, and which I shall never be able to recompose; and this has been done to one who never did, or imagined, you any harm. . . . In this business we are the sheep and you are the wolves. We will preserve our character and hope you will change yours. At all events we return you blessings for curses.

As the French Revolution grew bloodier, Priestley's position in England became more precarious. Both his esteem in the eyes of the

French revolutionaries and the fury of English conservatives reflect Priestley's impact. On August 26, 1792, the French legislature made some of the greatest advocates of liberty French citizens. Priestley headed the list, followed by Thomas Paine, Jeremy Bentham, William Wilberforce, and George Washington. The Revolution beheaded the king on January 21, 1793; England declared war on France in February. With friends and colleagues being transported to Australia for sedition, Priestley had no choice. On April 8, 1794, at the age of sixty-one, he left England for America, never to return.

In America, Priestley pushed on with his research and writing even as his health weakened. He performed a series of experiments to refute the claim by Erasmus Darwin and Jan Ingenhousz that organisms such as the green algae Priestley had studied appeared spontaneously. Priestley showed experimentally that some exposure to the atmosphere was always needed for the algae to appear. As Leeuwenhoek had done seventy years earlier, Priestley argued strongly that organisms could be generated only by similar organisms. He also completed the last four volumes of his *General History of the Christian Church*, followed by his *Index to the Bible*. He was dictating corrections an hour before he died. Earlier in his life Priestley had written, "Human happiness depends chiefly upon having some object to pursue, and upon the vigour with which our faculties are exerted in the pursuit." If that's the case, he died a happy man.

15

Humphry Davy, Intoxicated with Discovery

All the varieties of material substances may be resolved into a comparatively small number of bodies, which, as they are not capable of being decomposed, are considered in the present state of chemical knowledge as elements.

—*Humphry Davy, 1813*

Humphry Davy (1778–1829) was the scientific superstar of his day, with the crowds, the cachet, the groupies—and the ego—to match. A well-known portrait by Sir Thomas Lawrence captures it all—the brilliant young man at the top of his game, movie-star handsome, fashionably dressed and coiffed—a scientific Napoleon. There's no doubt his fame was justified, earned through brilliant work from his late teens through his early thirties. In those years he sketched out an original and insightful view of heat and light, discovered six new chemical elements, presented science with its first understanding of the electrical nature of chemical bonds, and offered the world its first anesthetic—a gift it spurned for forty-four years. Like some of the Romantic poets he partied with, he rode his intuition like a wild horse, worked in manic, feverish bursts, and as he described it himself, "burned out" at an early age.

Davy's childhood conveys the promise—and the edginess—of an isolated youth of great potential. He was an exuberant and imaginative child, far more interested in wandering the rocks and coves of Penzance than in attending school. His father, a woodcarver who lost what little money he made, died when Davy was sixteen. His mother, described as capable and attractive, supported her brood of five by running a millinery shop. After a self-described "dangerous year" of idleness preceding his father's death, Davy was apprenticed to a local apothecary-surgeon. At seventeen, Davy decided to get serious about his education. In the next two years he managed to teach himself something of theology, philosophy, several languages, and a great deal of science.

One of the books Davy devoured in his teens was Lavoisier's *Traite Elementaire de Chimie*. In it Davy came to face to face with his fate, and the result was explosive. Lavoisier had swept the venerable concept of phlogiston from its central role in chemistry. But he'd replaced it with a new, equally mysterious substance called caloric. Lavoisier argued correctly that Priestley's "dephlogisticated air" was oxygen, but insisted that oxygen was a compound of an unknown substance *plus* caloric. Nineteen-year-old Davy decided to take on the French master chemist. The result, after a few months of experimentation, was a twenty-thousand-word manifesto, "An Essay on Heat, Light, and the Combinations of Light," published in a collection of regional scientific works in 1799. Davy later distanced himself from those "infant speculations." Still, his essay sparkled with original and ingenious experiments, compelling lines of reasoning, and startling insights. Heat was not a substance, as Lavoisier claimed, but "repulsive motion" among the particles of a substance. The states of matter—solid, liquid, or gas—depended on the balance between attractive and repulsive motions. So hydrogen and nitrogen were not obliged to be gases; it might be possible to turn them into liquids or solids. Objects expand when heated not because they swell up with caloric, but because of increased repulsive motion of the particles within them. Light, Davy proposed, was simply another state of matter in which repulsive motion dominated, projecting the particles of light at enormous speed. He argued that "electric fluid" was a kind of condensed light, and that light readily combined with other substances; for example, those showing phosphorescence. He intuited that nerves carry electric fluid to the brain, and that muscles generate electric fluid when they contract. He argued that every change in perception or ideas must correspond to an actual change in the body. He proposed to measure and study those changes, with the aim of minimizing pain and increasing pleasure. "Thus would chemistry, in its connection with the laws of life, become the most sublime and important of all sciences."

Davy's essay greatly impressed Thomas Beddoes. He had recently founded the Pneumatic Institution at Clifton, where he hoped to use the new "airs" chemists were discovering to treat illness. He hired Davy, at nineteen, to head the laboratory. Davy had already experimented with small quantities of nitrous oxide, discovered by Priestley in 1774. He made an exploration of its properties his first project at Clifton. In addition to pinning down the chemical properties of the gas, Priestley studied its effects on animals and humans, characteristically using himself as a subject.

On April 17, 1799, Davy found that inhaling the gas produced striking behavioral and perceptual changes. In his laboratory notebook, his neat handwriting suddenly balloons exuberantly. "Davy and Newton," he scrawls in a moment of grandiosity. Sober again, he wrote, "The objects around me became dazzling, and my hearing more acute. Towards the last inspirations, the thrilling increased, the sense of muscular power became greater, and at last an irresistible propensity to action was indulged in." Not surprisingly, he continued his research. In July he inhaled nitrous while suffering the pain of an erupting wisdom tooth. To his surprise, the pain vanished, at least for a while. He was not so lucky inhaling other gases—he nearly killed himself breathing "water gas," which contains highly toxic carbon monoxide.

Davy described his ten months of intense research in an almost-600-page book on nitrous oxide, written in a few months and published before the end of 1799. The work was widely read and admired. In addition to detailing his chemical analyses of nitrous oxide and other nitrogen compounds, he discussed the painkilling effect of the gas, and argued that it could be used safely and "with great advantage" to block the pain of surgery. The medical establishment, to its discredit, took nearly forty-five years to implement his suggestion. Ironically, it took British society, alerted to the mind-bending effects of nitrous oxide by Samuel Taylor Coleridge and other literary admirers of Davy, no time at all to co-opt "laughing gas," for recreational use. It quickly became the rage at fashionable "frolics," the ecstasy raves of that time.

Davy had outgrown his position, and was soon recruited to another newly formed organization, the Royal Institution of Great Britain. Count Rumford, a leading scientist and humanitarian, had

SIR HUMPHREY DAVY.

Humphry Davy.

founded it to disseminate new and useful scientific knowledge. Davy started as an assistant lecturer, but was promoted to professor of chemistry within a year. He propelled the Royal Institution and himself into the limelight through his immensely popular lecture-demonstrations. The rich and powerful, along with aspiring scientists such as the young Michael Faraday, filled the Institution's auditorium to watch Davy's dramatic electrical and chemical demonstrations. Davy molded the Institution into a leading research center and kept it afloat financially through his lectures.

While Davy was still working on nitrous oxide, word flashed through scientific circles that an Italian researcher, Alessandro Volta, had created a continuous source of "electric fluid." Davy rushed to explore this new field. He built his own Voltaic pile — an early battery — and began a series of immensely productive experiments. As he studied electricity, he began to believe that it lay at the heart of chemistry. By 1806 he was convinced that charge is "an essential property of matter," and that it and what we would now call the chemical bond are identical.

If chemical bonds were electrical in nature, he realized, they could also be broken by electricity. "However strong the natural electrical energies of the elements of bodies may be," he wrote, "there is every probability of a limit to their strength; whereas the powers of our artificial instruments seem capable of indefinite increase." He realized that a powerful Voltaic pile might be just the tool he needed to break compounds down to their elementary components, or to prove that certain substances could not be split at all. Davy was the first to insist that the defining feature of a chemical element is that it cannot be decomposed in any chemical reaction.

His intuition proved right. On October 6, 1807, he used a battery made of 250 four-by-six-inch plates of zinc and copper to send a strong current through a small piece of moistened potash. "There was a violent effervescence at the upper surface;" he wrote, "at the lower, or negative surface . . . small globules having a high metallic lustre, and being precisely similar in visible characters to quicksilver, appeared, some of which burnt with explosion and bright flame, as soon as they were formed." He named the highly flammable substance potassium (later potassium). Davy was overjoyed as he watched the first globules of this never-before-seen metal appear. Although all he wrote in his notebook was "Capital experiment today," his brother described him leaping ecstatically around the room. Davy later said that he felt as though he'd realized the dreams of the alchemists.

Davy used every trick he knew, which included capitalizing on the outpouring of sympathy caused by a mysterious, eight-week illness he suffered in 1808, to raise money for more powerful batteries. He eventually built one with 2,000 plates, capable of sustaining a

four-inch arc far brighter than the Sun. (One spin-off of his work was
the carbon-arc lamp, the most powerful source of light until the
discovery of the laser.) His discovery of potassium was quickly followed by magnesium, calcium, strontium, and barium, all produced
from previously unbreakable compounds through electrolysis. In
1808, he added one more new substance, boron, to his bouquet of
new elements.

Chemists agree that Davy did his best work analyzing what we
now know to be the element chlorine. He was inspired by his distrust
of Lavoisier's theory of acids, which demanded that all acids contain
oxygen. The key, Davy thought, was muriatic acid, which yielded the
corrosive gas called oxymuriatic acid by Lavoisier. Davy attacked the
problem experimentally with every weapon he had—forming the gas
through a variety of reactions, and attempting again and again to derive oxygen from it or its compounds. He proved to his satisfaction
that the gas was composed only of hydrogen and an unknown substance, which he called chlorine from its greenish-yellow color. With
a combination of incisive experiments, precise observation and measurement, and rigorous logic, he proved that chlorine could not contain oxygen. Lavoisier's "oxymuriatic acid" was a misnomer, and
oxygen was not the key to acidity. Davy intuited instead that positive
and negative charges of the "particles" of different substances were
what determined their relative acidity or alkalinity.

In keeping with the goals of the Royal Institution, Davy was often
asked to interrupt his own researches to take on practical challenges.
In 1812 England suffered one of its increasingly frequent mine explosions, touched off by a miner's lamp. This one killed ninety-two men
and boys. Davy tackled the problem with his usual incisiveness. "It
will give me great satisfaction if my chemical knowledge can be of any
use in an enquiry so interesting to humanity," he wrote. He visited
mines, obtained samples of "firedamp," the explosive, methane-laced
gas that seeped into them, and experimentally determined the conditions that lead to explosions. He found that firedamp had a relatively
high ignition point, and that keeping the temperature of the gas below
that point could prevent explosions. Within months he devised an elegant way to do that—by surrounding the lamp flame with a metal
screen. His invention, the miner's safety lamp, revolutionized coal
mining. But, as with nitrous oxide, human nature provided an ironic
twist. The lamp did not reduce the number of miners killed, as Davy
hoped. Instead, it allowed mines to be dug into deeper and more dan-

gerous veins. His lamp may not have saved lives, but it did help fuel the Industrial Revolution.

Davy had completed most of his important scientific work by the age of thirty-three. In 1812 he was knighted, gave his last lecture at the Royal Institution, and married Jane Apreece, a rich and well-connected widow, all within a week. In the same year he published the first volume of his *Elements of Chemical Philosophy*. He never produced the rest of the promised work. The Swedish chemist Berzelius, who also carried out groundbreaking research on electrolysis, wrote that Davy's unfinished work could have moved chemistry a century ahead. Although Davy achieved further recognition, including presiding over the Royal Society from 1820 through 1827, and continued to advance the public understanding of science, he seemed to have lost some crucial component of his scientific prowess. His marriage was not a happy one, and he spent more and more of his time in society, traveling around Europe, and fishing. He had rocketed to fame, but now seemed to glide into obscurity. His health declined, and he eventually suffered a stroke. At the age of fifty, in Rome, he described himself as "a ruin amongst ruins." He died at fifty-one.

His influence, however, lives on. The historian of science David Knight emphasizes Davy's groundbreaking research in electrochemistry, but also his determination to utilize science to improve the human condition. "He was the apostle of applied science," Knight writes, "perceiving and urging that Britain's prosperity depended upon it, and in the safety lamp showing that knowledge was indeed power, and that science could ameliorate life."

16

Visionaries of the Computer

[T]he Analytic Engine weaves algebraic patterns
just as the Jacquard Loom weaves flowers and leaves.

—*Ada Lovelace*

The ascent up the hill of science is rugged and thorny,
and ill-fitted for the drapery of a petticoat.

—The Edinburgh Review, *1845*

"Man wrongs, and time avenges."

—*Charles Babbage, quoting Lord Byron*

It's a strange story, but true. In the middle of the nineteenth century the most brilliant British mathematician of his time set mathematics aside and spent the rest of his life and a great deal of money trying to build a computer out of crankshafts and gears. Too busy improving it to describe it adequately, he found his best interpreter in the beautiful, brainy, and star-crossed daughter of the infamous poet Lord Byron. Charles Babbage (1792–1871) never managed to build his computer. Ada Byron Lovelace (1815–1852) died young and under tragic circumstances. Yet, together they achieved something remarkable: at a time when steam engines were at the cutting edge of technology, he designed the Analytical Engine—a nuts-and-bolts machine that functioned like a modern computer, and she wrote the first program for it.

The son of well-off parents, Babbage never had to work for a living. He was a smart and curious child who always seemed to know more than his teachers. By the time he entered Cambridge in 1810 he was clearly brilliant, but also independent, confident, and sociable. He cut classes to go sailing, stayed up all night playing cards, and still led his class. He and a few fellow students created the Analytic Society, whose goal was to bring backwater British mathematics up to European standards. They eventually succeeded, as was recognized years later with Babbage's appointment to the Lucasian Chair of Mathematics, once held by Newton (and today by Stephen Hawking).

After Cambridge, Babbage married and moved to London. He turned out original work on the theory of functions, and continued his efforts to reform not just British mathematics, but science as well. He became an outspoken and effective critic of the Royal Society, and helped found new scientific groups, including the Astronomical Society and the British Association for the Advancement of Science. But he soon became inspired—or obsessed—with a revolutionary idea, to bring the power of the machine to bear on the mathematician's bane, calculation.

The idea first came to Babbage while he was still at Cambridge. He and a friend, the future astronomer John Herschel, were comparing calculations for a mathematical table, and finding many errors. Wouldn't it be possible, Babbage wondered, to grind out such mind-numbing computations with some kind of calculating engine? At the time he simply convinced himself that the idea could work. But, in 1821 he sidelined most of his other interests to design and build a working calculator, which he called the "Difference Engine." He envisioned it as a machine that would calculate to many decimal places any mathematical function that could be expressed as the sum of up to six successive powers of a variable.

The problem of inadequate or inaccurate tables was a serious one in many scientific and practical areas, including navigation, which was vital to England's prosperity. Babbage utilized his status and connections to do something unprecedented at the time—win government support for his project. Starting with an initial £1,500, the British government eventually invested £17,000 towards the Difference Engine.

Unfortunately, Babbage was a perfectionist. Before he could build the Difference Engine, he insisted on designing and fabricating precision tools and machines to make its parts. He continually changed and improved on his plans. To complicate matters, he was a frequent and outspoken critic of influential scientists and decision makers. The British Government came to see Babbage as a problem and his Difference Engine as a bottomless pit. One result was that he wasted a dozen years lobbying, badgering, and pleading with the government just to find out if the project was dead or alive. In 1842, Astronomer Royal George Airy pronounced the official verdict on Babbage's machines — "worthless."

The only tangible result was a working model of the Difference Engine. Babbage kept it for many years to amuse the many visitors to his studios, including Ada Lovelace. The Engine currently resides in the Science Museum of London and reportedly still works. Although he was embittered toward the government, Babbage was willing to let the Difference Engine languish. He had long since moved beyond it to envision a far more flexible and powerful machine, the Analytical Engine.

What Babbage realized was that many Difference Engines could be controlled by a single mechanism. The mechanism he designed was based on punched cards like those used to control the Jacquard loom. Rather than designating which threads to raise and lower to weave a pattern, Babbage's punched cards would provide step-by-step instructions to carry out calculations of any degree of complexity. He set out to design a machine to accomplish this. Babbage eventually produced hundreds of pages of exquisite mechanical drawings with matching functional diagrams in a notational language of his own invention. Experts say that the Analytic Engine may well be the most complex system ever designed by one person.

By the end of 1837 — sixteen years after first being bitten by the computing bug — Babbage had pieced together the essential ingredients of a fully programmable computer:

1. Input: He planned to use punched cards to move numbers and, more importantly, instructions, into the machine.
2. Memory, which he called the "store": Part of the store was inside the machine, where he planned to hold up to 1,000 50-digit numbers. Information could also be exported to punched cards for later retrieval and use.
3. A central processor or "mill": This registered numbers and instructions, translated them into appropriate internal settings,

and operated on them as specified, synchronized, and se-
quenced by a clock.
4. Output: This could be in the form of printed tables, graphs, or
   punched cards.

Babbage also invented mechanical ways to implement two other
vital functions of a computer—repetitive loops or subroutines, and
branching or conditional choices. The basic functions he discovered—and found ways to carry out with levers and gears—form the
conceptual basis of every computer operating today.

Babbage first met Ada Lovelace when he was forty and she was seventeen. She was fascinated by his brilliance and by the beautiful machine he had created. Ada was the only legitimate child of the poet
George Gordon, Lord Byron, who was described by a contemporary as
"mad, bad and dangerous to know." Ada's mother, Anne Milbanke, left
Byron a month after Ada was born, following a tumultuous and violent
year of marriage. From birth, Ada's life was shaped by her mother's
steely determination to stamp out any hint of Byron in her daughter.
Lady Byron set out to do this with a fistful of disciplinary techniques
that would be seen as viciously abusive today—such as forcing Ada to
lie perfectly still for hours, and locking her in a closet if she moved as
much as a finger. When Ada was a bit older, Lady Byron forced her to
study mathematics. She saw mathematics as the surest way to discipline
the unruly spirit lurking in her lovely but unloved daughter.

At twenty, Ada married a gentleman, Lord Lovelace, and soon produced three children. Ada's mother approved of the marriage, since
she found the bland Lovelace easy to control. In her mid-twenties,
however, Ada made a move that threatened her mother's domination.
Rather than let mathematics quench her passion, she turned mathematics into a passion. Inspired by Babbage, she made mathematics her
religion, and herself a divine acolyte. She would be a "prophetess" of
mathematics, the "*vocal* organ" of a mathematical deity. "I am simply
the *instrument* for the divine purpose to act on & thro'," she wrote to
her mother, words that must have made Lady Byron curse Ada's father
yet again. Lovelace's wealth and social standing allowed Ada to devote
most of her time to her studies.

On the suggestion of the scientist Charles Wheatstone, Ada translated into English a description of the Analytical Engine by an Italian
mathematician, Louis Menebrea. By that time, Babbage and Lovelace
had formed a complex, teasing friendship. Lovelace had learned a

great deal about the Analytical Engine from him. Babbage, who never missed an opportunity to advance the cause of his Engine, suggested to Ada that she supplement Menebrea's description with notes of her own. Lovelace threw herself into the project. In the process she produced a unique vision of the Analytical Engine's potentials.

In her "Notes," Ada described for the first time what the Analytical Engine could do. She pointed out that mathematics was the language of science, the only way to analyze natural relationships and processes. By speeding up computation, she wrote, the Engine would be an immensely powerful scientific tool. She also saw that a programmable machine could do more than crunch numbers; it could manipulate any kind of symbols. The Analytical Engine could solve equations, she realized, and it could even compose music. To show how the machine would work, she wrote the world's first computer program—an algorithm to calculate a numerical sequence discovered by Bernoulli.

Ada's commentary appeared in *Taylor's Scientific Memoirs* in 1843, identified only by her initials. They made a very favorable impression, and they won Ada a bit of the recognition she desperately needed. More than a century later, Ada's Notes would win her a unique kind of immortality—ADA, the high-level computer language used by the U.S. Department of Defense, was named after her, the world's first programmer.

Babbage never built the Analytical Engine, although his son later assembled a small version of the "mill." But his ideas provided the conceptual foundation, nearly a century later, for the first electronic computers. In the end, Babbage outlived his critics, but also his era. He died at the age of seventy-eight, a cantankerous and embittered man.

The last years of Ada's life were even more tragic. Notorious from birth because of her father, her soul the prize in her mother's vendetta against him, she desperately sought to be worthy on her own. "Far be it from *me* to disclaim the influence of *ambition* and *fame*," she wrote to Babbage. "No living soul ever was more imbued with it than myself." Sadly, with the exception of her essay on the Analytical Engine, she never lived up to her grandiose dreams. In a parody of her gamble for fame on the Analytical Engine, she began to gamble on horses. She became financially and romantically entangled with an unscrupulous man, John Crosse. He tried to blackmail her, which forced Ada to confess her financial losses and her affair to her husband. Ada's "fall" made her even more vulnerable to her mother. Ada became ill, which allowed Lady Byron to reestablish a stranglehold over her. Ada begged for opium to ease the pain from her uterine cancer. Her mother would not allow it, on the grounds that suffering would help her to repent. Ada asked to see her old friend Babbage. Her mother would not allow that either; he had helped Ada win a few moments of independence. Lady Byron wrote to a friend to describe the profound gratification she felt when Ada, in agony, finally declared that the only reason she might want to live would be to devote herself to her mother.

In contrast, Babbage seems never to have doubted himself or his ideas. "As soon as an Analytical Engine exists," he predicted, "it will necessarily guide the future course of science." His prediction is finally being realized today. Computers have grown powerful enough to allow scientists to model enormously complex phenomena such as the Earth's climate; the evolution of stars, planets, and galaxies; the genomes of plants, animals, and humans; and the convoluted folding of the protein molecules of which we are made. Without computers, little of today's science or technology would exist. Still, as Ada might point out, they have not even begun to sort out the intricacies of the human heart.

17

Darwin's Great Truth

I happened to read for amusement Malthus on *Population,* and being very well prepared to appreciate the struggle for existence which everywhere goes on . . . it at once struck me that under these circumstances favourable variations would tend to be preserved, and unfavourable ones to be destroyed. The result of this would be the formation of new species. Here then I had a theory by which to work.

—*Charles Darwin, October 1838*

He found a great truth, trodden underfoot, reviled by bigots, and ridiculed by all the world.

—*Thomas Huxley, 1882*

Charles Darwin (1809–1882) was one of those scientists whose future talent was invisible to many of the adults around him. The headmaster at Shrewsbury school scolded him for wasting his time, and his frustrated farther, a successful physician, predicted he'd be a disgrace to himself and his family. Darwin later wrote, "The school as a means of education to me was simply a blank." Nature, however, fascinated him—he was an avid collector of rocks, plants, and insects. He loved observing the habits of birds but, until his mid-twenties, loved hunting them even more.

At sixteen, his father sent him to study medicine at Edinburgh University. Always sensitive to suffering, he was horrified to see two patients, one of them a child, undergo surgery without anesthesia. "The two cases haunted me for many a long year," he later wrote. He convinced his father to let him quit after two years. With the help of a tutor, he got into Cambridge, supposedly to study theology. Darwin the budding naturalist spent his days collecting beetles, while Darwin the sportsman spent nights drinking and playing cards with his high-spirited friends. Darwin the theologian never appeared. Still, he made it through, graduating in 1831.

By his own estimation, Darwin did not learn much at Cambridge, but he did form a deep friendship with one of his professors, the naturalist John Stevens Henslow. Darwin spent hours with this open-minded and warm-hearted scientist who was the first to sense his potential. It was Henslow who interested Darwin in natural history, who gave him the idea that he might do more with his life than collect beetles and shoot birds, and whose connections won Darwin his life-changing berth as unpaid naturalist and companion to Captain FitzRoy aboard HMS *Beagle*.

Darwin's five years aboard the *Beagle* transformed him. He brought with him a copy of Charles Lyell's just-published *Principles of Geology*, and was soon applying Lyell's dynamic understanding of geological processes to the fossils and formations at the *Beagle's* ports of call. Darwin explained for the first time how coral reefs and atolls form, discovered the fossilized bones of giant sloths and armadillos, and observed how the Chilean earthquake of 1835, which destroyed the town of Concepción, had raised the surrounding land by many feet—a powerful vindication of Lyell's theories. These observations alone would have made his scientific reputation, but throughout the long voyage of discovery he was cataloging and studying every plant and animal he could find. At some point in the journey he was no longer simply collecting novel specimens; he had started to ask deep questions and seek their answers: How did these species come to be where they were? How did they develop the unique features that so elegantly fit them into their places in nature? How do species come into being at all?

The long voyage of the *Beagle* transformed Darwin's personal life as well. As a young man he'd always been ready for a geological outing, a hunting trip, or an evening with friends. But after returning to England he began to suffer from a mysterious illness that would cause him great discomfort and long periods of disability throughout his life. The doctors he consulted never found the cause of his digestive problems, chills, and malaise. In 1990, John Bowlby, a psychiatrist who spent his life studying the effects of early separation and loss, speculated that Darwin suffered from a severe anxiety disorder stemming

Charles Darwin.

from the loss of his mother. She was often ill, and died when he was eight years old. Recently Saul Adler, an Israeli parasitologist, suggested that Darwin's symptoms match those of Chagas' disease, caused by a parasite Darwin was exposed to in Argentina. Whatever its source, his illness led Darwin and his wife to move to a quiet country estate where they lived and raised their children in near seclusion. Darwin even avoided visits with fellow scientists—the excitement caused fits of shivering and vomiting. He lamented the years lost to his illness, but his sequestered lifestyle seems to have given him the time he needed to build his grand theories.

Darwin was not the first to think about evolution. His grandfather, Erasmus Darwin, had written about it in his book *Zoonomia*. The elder Darwin, like the French naturalist Lamarck, believed that organisms passed on to their offspring qualities they had acquired during their lives—a gazelle that spent its life trying to outrun predators would produce even swifter offspring. But the dominant view of Darwin's time was deeply antievolutionary. Most people, including most scientists, believed that existing species had been created one-by-one just a few thousand years ago, as taught in the Bible.

The theory of evolution did not spring full-blown into Darwin's mind. He puzzled for years over the distribution of the plants and animals that he studied during the voyage of the *Beagle*. It wasn't until nine months after his stay in the Galapagos Islands that he first suspected that the obviously related, but slightly different kinds of birds on neighboring islands might stem from recent evolution, and so might "undermine the stability of evolution." A year after Darwin's return to England, John Gould, the ornithologist classifying his specimens, confirmed that the mockingbirds Darwin had found on neighboring islands represented distinct species. That convinced Darwin that new species could form when populations became geographically separated from one another, and that similar species could derive from a common ancestor.

Still, Darwin remained mystified by what might cause evolution. He considered and rejected dozens of ideas. Natural selection, the engine of evolution, did not become clear to him for another year and a half. The spark that let Darwin fit the pieces together was struck by Thomas Malthus's grim essay about what we now call population pressure. Malthus was writing about human populations, but Darwin realized that every species produces far more offspring than can survive. He was the first to see that nature does not thin the ranks of a species at random. Individual plants and animals that are even slightly more fit are more likely to survive and reproduce, while less fit individuals are culled from the breeding population. Just as farmers modify their herds and crops by deliberate selection and breeding, nature shapes existing species and eventually forms new ones by selecting those organisms that are best adapted to current conditions to seed the next generation. Natural selection is the sieve, and population pressure is the force pushing each generation through it.

In that moment of vision—one that he had spent years preparing for—Darwin recognized that all species, existing and extinct, had evolved over time from earlier forms, and that all the intricate adaptations of living things, which until then had been attributed to God's handiwork, were products of the impersonal force of natural selection working slowly but inevitably over vast spans of time.

Darwin's crucial insight—natural selection—was brand new. Unlike many great scientific ideas, it had not been glimpsed by the ancient Greeks, debated by philosophers, or proposed by an earlier scientist. As Thomas Huxley, Darwin's eloquent advocate, wrote, "The suggestion that new species may result from the selective action of external conditions upon the variations from the specific type which individuals present … is as wholly unknown to the historian of scientific ideas as it was to biological specialists before 1858. But that suggestion is the central idea of the *Origin of Species* and contains the quintessence of Darwinism."

Darwin experienced that insight in 1838, but he published *On the Origin of Species* in 1858. Nobody knows exactly why he waited twenty years to present his discovery to the world. We do know that the few scientific friends to whom he confided his thoughts begged him to publish, and that his hand was finally forced when the naturalist Alfred Russel Wallace sent Darwin a paper he had just written that duplicated the key features of evolution by natural selection. Like Copernicus, Darwin was fully aware that his idea threatened the existing worldview. And, like Copernicus, he spent years gathering every fact and every argument that could support his new system. If Wallace had not stumbled into the territory Darwin had discovered, it's possible that Darwin, like Copernicus, might have waited until he was dying to publish his work.

At the end of *The Origin of Species*, Darwin hinted that the theory would shed light on human evolution. Darwin braved those turbulent waters in his 1871 book, *The Descent of Man*. In it he presented evidence supporting the close relationship between humans and apes. He detailed many physical similarities, found parallels throughout the developmental journey from embryo through reproductive maturity, and even found analogs of human emotions and behavior patterns in apes. He showed how all of these features could have evolved by natural selection as modern apes and humans differentiated from a shared ancestor. Of all Darwin's ideas, this one met the fiercest resistance.

With the exception of Darwin's closest supporters, hardly anyone liked the idea that humans and apes sprang from a common ancestor. Theologians were especially appalled. Darwin's theory flew in the face of the Biblical assertion that God created man in his own image. To theologians, mankind was unique, the only animal with a soul, little lower than the angels. To suggest otherwise seemed to threaten the very definition of humanity, the basis of morality, in short, their view of the world.

Darwin's mechanistic explanation of adaptation also robbed believers of one of their most powerful proofs of the existence of God, the argument by design. Like today's creationists, theologians of Darwin's time argued that complex organs such as the eye, not to mention organisms whose every feature suited them to their environments, proved the existence of a Great Designer. The Reverend William Paley had written in 1802 that just as a watch implied the existence of a watchmaker, nature's intricate works implied an all-knowing Designer. The impersonal laws of nature might direct the path of a planet, but only God could fashion living things, so beautifully differentiated and adapted. Darwin's theory made the Watchmaker superfluous; it was not God but natural selection, winnowing away over geological time, that had fashioned the eye of the eagle, the legs of the gazelle, and the claws of the tiger.

It's sometimes said that Darwinian evolution is based on a circular argument. If fitness is measured in terms of survival and successful reproduction, then "survival of the fittest" has no meaning. Darwin, if not his critics, understood that natural selection did not make plants or animals better in any absolute sense; it simply shaped a species to thrive in its particular environment and lifestyle. Darwin discarded the human chauvinism symbolized by the universally accepted ladder of creation, or Great Chain of Being, which ranked all organisms from the simplest on up to the top rung—mankind. Darwin was the first to realize that every living species can trace its descent to a single source—we are all twigs on a densely branching tree of life. To him, species were not higher or lower—they had simply adapted to different environments and ways of life.

It was also in *The Descent of Man* that Darwin first focused on an extremely important component of the evolutionary engine—sexual selection. He was the first to realize that since evolution hinges on successful reproduction, features that make successful mating more likely will evolve particularly quickly and dramatically. Darwin realized that sexual selection explained many of nature's most striking displays—a stag's massive antlers, a peacock's exquisite tail, the fierce battles between competing males—as well as our own secondary sexual features such as beards and breasts. Today it's known that sexual selection is based not just on male competition, but on female choice, and that it shapes every detail of reproduction, from the shape, size, and color of reproductive organs to the intricacies of mating behavior.

Perhaps the strongest test of a theory is its ability to make successful predictions. Darwin's theory led him to believe that flowering plants and the insects that pollinate them had evolved together. He studied many plants whose flowers could only be fertilized by a particular kind of wasp, bee, or moth. So when he learned of the Christmas Star orchids on Madagascar that store nectar at the bottom of a foot-long tube, he predicted the existence of a moth with a proboscis that could unfurl to the same length. Experts ridiculed the idea until, some forty years later, a moth equipped with just such an improbable organ was discovered. Modern evolutionary theory, using sophisticated mathematics to calculate the shifting frequency of genes within populations, has produced far more specific predictions; for example, the high level of relatedness that underlies the apparent altruism seen in social insects.

From the start, Darwin's theory stirred general hostility, resistance, and ridicule. But it also faced serious scientific challenges. The most significant threat came from an unexpected source—the physicists of his day. Armed with the gleaming weapon of the laws of thermodynamics, Lord Kelvin and other leading physicists slashed away at the necessary backdrop of Darwinian evolution—time. Kelvin calculated that the sun must have been far hotter not too long ago and had been cooling rapidly ever since. The Earth could not be more than 30 million years old, he showed, the time it would have taken to cool from a molten state. That was more than the 6,000 years granted by the Bible, but far less than natural selection required. Most geologists fell in line with Kelvin's seemingly irrefutable numbers. Darwin and his disciples struggled unsuccessfully to squeeze what we now know to be more than 3 billion years of evolution into Kelvin's straitjacket. In 1871, Darwin wrote to Wallace, "I have not as yet been able to digest the fundamental notion of the shortened age of the sun and earth." By 1873, Kelvin contemptuously described Darwin's theory as utterly futile.

It turned out to be the fuzzy-headed biologists, not the clear-thinking physicists, who were right. Unfortunately, Darwin died fifteen years before Becquerel discovered radioactivity. By the turn of the century, the Curies, followed by Ernest Rutherford, showed that radioactive elements within the Earth were a powerful source of energy that Kelvin had not dreamed of. They made his chilling calculations irrelevant, and gave natural selection the time it needed to do its work.

One controversy that is still being debated is the central role Darwin gave to gradual change. He knew that the fossil record often seems to tell a different story—of long periods of stability marked off by dramatic changes. Darwin went to great lengths to point out, correctly, the incompleteness of the fossil record. He predicted, sometimes correctly, that intermediate forms would eventually be found to fill in a history of slow, continuous change. But today most biologists favor the theory of Punctuated Equilibrium proposed by Niles Eldredge and Stephen Jay Gould in 1977. They argue that evolution typically moves in surges—long periods of relative stability are punctuated by sudden changes in the nature and mix of species. We now know that some of the greatest punctuations—in which a large percentage of existing species disappeared, eventually to be replaced by new species—were caused by catastrophes such as a huge asteroid slamming into the Earth.

Like today's genetically modified crops, Darwin's ideas of competition and natural selection refused to stay in their proper field. Theologians were the first to react to the incursion into their territory, starting a battle that is still being fought by Creationists today. Despite the theologians, and against Darwin's own wishes, Darwinian ideas of all-pervasive competition and "survival of the fittest" soon permeated almost every area of human thought—anthropology, economics, and politics, to name a few. In many cases, they were appropriated to justify discrimination, racism, and cutthroat capitalism. Darwin, who hated slavery and any form of cruelty, would have been appalled. Still, he might have been intrigued by some generalizations of his ideas. Today, for example, natural selection is used to try to explain the origin of life itself in a primordial world of competing, self-replicating molecules. And some cosmologists argue that our stable and livable cosmos was selected from a myriad of universes, most of which had the bad luck to be spawned with less congenial packages of particles and forces.

Darwin loved nature's abundance of forms and features. As he studied them, he found within them the workings of a simple process—yet one so pervasive, persistent, and powerful as to account for the existence and characteristics of all living things, including us. Perhaps the best assessment of the impact of Darwin's ideas was made by the geneticist Theodosius Dobzhansky, who said simply, "Nothing in biology makes sense except in the light of evolution." But nobody has summed up his vision more beautifully than Darwin himself:

> There is a grandeur in this view of life, with its several powers, having been originally breathed into a few forms or into one; and that, whilst this planet has gone cycling on according to the fixed law of gravity, from so simple a beginning, endless forms most beautiful and most wonderful have been, and are being evolved.

18

A Genius
in the Garden

It requires indeed some courage to undertake a labor of such far-reaching
extent; this appears, however, to be the only right way by which we can finally
reach the solution of a question the importance of which cannot be overesti-
mated in connection with the history of the evolution of organic forms.

—*Gregor Mendel*

Like the humble garden peas he studied, Gregor Mendel's modesty
hid a remarkable set of qualities. Today no one doubts his brilliance, yet in some ways he was surprisingly limited. His religious
order sent him to college to train as a teacher, and he taught for years
at the local high school. But Mendel (1822–1884) was never able to
pass the examinations for his teaching credentials—the examiner
complained that Mendel lacked insight and clarity. Yet, working alone
and in his spare time, Mendel carried out an incisive series of experiments that unlocked the secrets of heredity—three decades before
anyone else was prepared even to understand what he had found. He
saw through nature's bewildering variety of colors, forms, and textures
to discover the fundamental units of heredity and the elegant laws that
govern them. Mendel was an unpretentious man, a roly-poly, twinkle-
eyed cleric who was loved by his students and peers. He met his many
responsibilities with patience, diligence, and notable good humor. Yet
his intellectual audacity was breathtaking. Within his little garden of
peas he sought and found some of nature's deepest secrets.

Mendel is sometimes depicted as an amateur who stumbled upon his discoveries. Nothing could be further from the truth. Despite his impoverished background as the son of a peasant farmer in rural Austria, he received a solid scientific education. His parents stretched their resources to the breaking point to send him to a gymnasium, or high school, miles from home, and his sister sacrificed part of her dowry to keep him in school. Several teachers saw his potential and helped out at critical points. His physics professor at the Philosophical Institute at Olmütz recommended Mendel to the abbot of the Augustinian Monastery of St. Thomas, in Brünn Moravia (now Brno, Czech Republic), in 1843.

The monastery was a center of intellectual life in Moravia. Its abbot, Cyril Napp, encouraged his monks to pursue scientific or artistic interests in addition to their monastic duties. Mendel studied theology and was ordained as a priest in 1848. He found ministering to the sick too disturbing, and was sent instead to teach at the high school in nearby Znaim. To formalize that position, he took an examination to become a certified teacher—and failed. Ironically, he earned adequate scores in physics and mathematics, but not in biology. Still, he was such a good teacher that the abbot sent him to study at Vienna University, where he acquired two years of first-class scientific training. The physics and chemistry he studied showed him that complex phenomena could be reduced to simpler units and explained by the working of scientific laws. He did well in his classes. Yet when Mendel took the teacher's examination a second time, in 1856, he failed again.

In the same year, Mendel began to cultivate a little strip of the monastery garden in his spare time. His far-from-modest aim was to uncover the laws of heredity. There's no doubt that he fully understood the importance of what he was doing. In his now-famous paper, published after a decade of experimentation, he makes it clear that he believed that there were simple and understandable laws underlying heredity, and had set out to find them. He was even aware that his work might help clarify the question of the origin of species. His research was anything but amateurish; Mendel closed in on his scientific prey through a series of incisively designed and keenly analyzed experiments. And he didn't stop there—he went on to restate his findings mathematically, used his mathematical model to make predictions, and validated those predictions in further experiments. It was a scientific tour de force.

Mendel knew that botanists were trying to bring order to the complicated and mysterious question of heredity. It was obvious that the offspring of plants and animals resembled their parents in some ways and differed from them in others. Most scientists believed that parental characteristics appeared in mixed or blended forms in their offspring. Nobody knew just what was passed from parent to offspring, nor how that transmission took place. Biologists were still arguing about whether plants reproduced sexually. Botanists had bred large numbers of plants but had only made qualitative observations. They recognized, for example, that most hybrids—the offspring of dissimilar parents—did not breed true, that the offspring of hybrids resembled the original parent plants, and that species differed greatly in how many generations it took to create a stable hybrid.

Mendel started by searching for a suitable plant to study. His choice, the common garden pea, turned out to be ideal. But he didn't stumble on it by luck. He searched for a species that met three criteria: he wanted a plant with stable varieties distinguished by a variety of characteristics; he needed to be able to control pollination; and the hybrids and their offspring had to be fertile. Even after settling on peas, he experimented with thirty-four varieties before choosing the ones he would work with. Similarly, he considered a large number of traits before picking seven that he could track reliably, generation after generation:

1. Shape of the (dry) seeds: round and smooth vs. angular and wrinkled
2. Color of the interior of the ripe seeds: yellow-orange vs. green
3. Color of the seed coat: white vs. gray, gray-brown, or brown
4. Shape of the seed pods: smooth vs. indented
5. Color of the unripe pods: shades of green vs. "vividly yellow"
6. Position of the flowers: along the stem vs. clustered at the tip of the stem
7. Length of the stem: longer than 6 feet vs. shorter than 1½ feet

Scientists and historians have remarked on Mendel's incredible luck. We now know that peas have just seven chromosomes. Had he chosen eight traits, two of them would have been inherited together to some degree, clouding his results. And even with seven traits, Mendel could easily have been hampered by several that happened to be linked. The traits Mendel chose turned out to be inherited independently of each other. At the very least, Mendel's intuition was working overtime. More likely, years of patient observation had led him to the seven best characteristics to study.

Next, Mendel crossbred his plants systematically and in large numbers. Over the years, he bred and studied over 12,000 plants. The crucial step was to fertilize each flower by hand with pollen from a different variety, then protect each flower from accidental pollination with a tiny cloth cap. In the first generation, Mendel found that rather than blended traits, certain characteristics always appeared while others disappeared. For example, all the offspring of a cross between long-stemmed and short-stemmed parents turned out to have long stems. He called that the dominant characteristic. This was a major finding in itself, disproving the traditional belief that hybrids inherited a blend of traits.

Mendel's next round of experiments gave him the key he needed. When he bred the hybrids, traits that had vanished in the first generation magically reappeared. And when he counted the long- and short-stemmed plants, the smooth or wrinkly seeds, the green or yellow pods, he found something truly remarkable—their numbers clustered around a simple ratio. Of 1,064 plants, 787 had long stems. Of 7,324 seeds, 5,474 were round. Of 580 pods, 428 were green. The ratios varied between 2.84:1 and 3.15:1, but they averaged 2.98:1. In other words, out of every four second-generation plants, three showed the dominant trait, while one showed the contrasting trait, which Mendel labeled recessive.

At that point Mendel was in a position to piece together the first comprehensive theory of heredity. He deduced that some material unit within both egg and sperm carried specific traits from generation to generation. He explained the disappearance and reappearance of recessive features in the simplest way possible—by assuming that the units of heredity appear in two forms, with one dominating the other. If a plant inherits either one or two dominant forms of a characteristic, it will show that trait. Only if it inherits two recessive traits will it show the recessive quality. With the added assumption that the units of heredity are reshuffled at random in each generation, Mendel was not only able to explain what he had seen in the first two generations, but also to predict the ratios of dominant and recessive traits in subsequent generations.

Mendel spent the next years checking out his predictions through six generations of plants. The theory worked perfectly. Although he cautioned that his findings needed to be replicated by others and ex-

Gregor Mendel.

tended to different species, he made it clear that he thought they applied to all living things. "In the meantime," he wrote, "we may assume that in material points an essential difference can scarcely occur, since the unity in the developmental plan of organic life is beyond question." A half century later the units of heredity that Mendel conceived would come to be called genes. Nearly a century later, Watson and Crick would specify the molecular structure of genes in the double helix of DNA. And today, molecular biologists can read, cut, and paste the molecular code almost at will. They now have the capability to modify organisms, diagnose and cure diseases, and, eventually, reshape the living world, including mankind.

Mendel published his findings in 1865, in *The Transactions of the Brünn Society for the Study of Natural Science*, under the unassuming title "Experiments with Plant Hybrids." With its publication, as Jacob Bronowski puts it, Mendel "achieved instant oblivion." Admittedly, the journal was an obscure one, but it was sent to 120 learned societies throughout Europe, and Mendel mailed reprints to forty leading biologists, including Charles Darwin. We know that Darwin didn't read it—his copy was found unopened among his papers. But even those who did, such as the eminent botanist Karl von Nägeli at the University of Munich, did not grasp its significance. Nägeli, at least, responded to Mendel. But he sent Mendel down a blind alley by suggesting *hieracium*, or hawkweed, as the next plant for Mendel to study. Hawkweed reproduces asexually, so it was useless for Mendel's purposes. Biologists continued for more than thirty years to think in terms of the mixing and blending of parental traits, blind to the simple and precise laws of heredity Mendel had tried to show them.

One measure of Mendel's unique genius is that Darwin also spent years experimenting with crossbreeding in several different species. Certainly nobody was more motivated than Darwin to discover the mechanism of heredity, without which his theory of evolution lacked a foundation. Yet even he failed to glimpse what Mendel saw so clearly.

Two years after Mendel published his groundbreaking work, he was elected abbot of his monastery. Although he wanted to continue his research, the new role demanded most of his time for the rest of his life. He remained interested in science, breeding new varieties of plants (and some highly productive but notoriously bad-tempered bees), replanting an eroded mountainside, and studying astronomy and meteorology. But he never again found the time to return to his work on heredity. He died at the age of sixty-three, admired by those who knew him, but unnoticed by the world of science. As if to guarantee his obscurity, his papers and letters were burned after his death.

It was not until 1900 that science finally caught up with him. Three European researchers, Hugo de Vries, Carl Correns, and Erich von Seysenegg independently carried out similar experiments and drew conclusions that echoed Mendel's work. De Vries was the first to rediscover Mendel's obscure paper and bring his work out of the shadows. Correns and von Seysenegg soon joined him in giving Mendel credit for his fundamental discoveries. So the dawn of the twentieth century marked the belated birth of the science of genetics. A century later, having decoded the entire human genome, we now know that from the seed Mendel planted would spring nothing less than a complete understanding of the tree of life itself.

19

Mendeleev Charts the Elements

It is the function of science to discover the existence of a general
reign or order in nature and to find the causes governing this order.
And this refers in equal measure to the relations of man—social
and political—and the entire universe as a whole.

—*Dmitri Mendeleev*

Mendeleev (1834–1907) was a driven man. In addition to producing his greatest work—the periodic law and periodic table of the elements—he researched, lectured, and wrote at a ferocious pace. A fine-print list of his published works takes up ten pages. Chemistry was the heart of his work, but he also played a major role in the economic development of Russia by modernizing that nation's weights and measures and through his advocacy of improved mining, manufacturing, agriculture, and trade. He fought ignorance and mysticism by reforming education and opening the sciences to women, and helped found and head the Russian Chemical Society. He habitually worked day and night, keeping himself going with a mixture of drive, determination, and strong Russian tea.

The source of his determination is no mystery. Mendeleev was molded, inspired, and set irreversibly on track by his mother, Maria Dmitrievna. She bore fourteen children, eight of whom survived their childhood in Siberia. Dmitri was the youngest and her favorite. Maria resolved to see his promise realized at any cost. When her husband Ivan, a teacher of Russian literature, went blind, Maria supported the family by managing a local glassworks. After the factory burned down, Maria sold their few possessions, packed fourteen-year-old Dmitri and his sister into a wagon, and set off for Moscow, where she hoped to get him admitted to the university to study science. When the university refused even to let him apply, she pushed on to St. Petersburg. Dmitri was not allowed to apply to the university there either, but he was accepted at the Main Pedagogical Institute. As if her life's work was done, Maria soon died. In the preface to one of his many books, Mendeleev wrote, "She instructed by example, corrected with love, and in order to give him to the cause of science she left Siberia with him, so spending her last resources and strength. When dying, she said, 'Be careful of illusions. Work. Search for divine and scientific truth.' ... Dmitri Mendeleev regards as sacred a mother's dying words."

With his mother's words echoing in his ears, plus his evident talents, Mendeleev became a gold-medal–winning student, tackled his master's and doctoral research at the University of St. Petersburg with fierce energy, and quickly emerged as one of Russia's leading scientists. In 1860, the Russian government chose him to attend the First International Chemical Congress, in Karlsruhe, Germany. The Congress catalyzed chemistry. It brought the world's leading chemists face to face to hammer out agreements on basic concepts, including such crucial issues as how to determine atomic weights. It was Mendeleev, the driven, wild-haired young man from Siberia—not J. A. R. Newlands of England, B. de Chancourtois of France, or Lothar Meyer of Germany— who left Karlsruhe charged with everything it would take to discover the periodic law and create the periodic table.

The idea that matter is composed of atoms dates back to the ancient Greeks. But only in the decades leading up to Mendeleev's discovery had chemists sorted out elements from compounds, determined the characteristics of about sixty elements, and made progress toward understanding atoms and molecules. They realized that some of the elements shared certain qualities. Iron, cobalt, and nickel, for example, were similar to each other, but very different from the so-called halogens, such as fluorine, chlorine, and bromine. A few chemists, including Newlands and Meyer, had seen signs of a deeper order hidden within the elements. In 1866, Newlands had proposed his "Law of Octaves," noting that when the elements were listed by increasing atomic weight, every eighth element tended to have similar properties. But most chemists remained skeptical. One asked Newlands if he'd also tried listing the elements alphabetically.

Newlands and the few other chemists searching for unity among the elements proceeded cautiously. Mendeleev, however, pushed on. He wrote to scientists across Europe to pull together the latest data on the elements and their compounds. Like Newlands, de Chancourtois, and Meyer, he believed that atomic weight—the mass of an atom compared to hydrogen—held the key to the properties of the elements. But in addition to the best estimates of atomic weights, Mendeleev also collected and collated other properties of the elements—their specific weights and volumes, the temperatures at which they solidify or vaporize, and the kinds of compounds they form.

Mendeleev mulled over his growing hoard of data for years, sorting the elements by weight; by family resemblance; by the way they combined with hydrogen, carbon, and oxygen; by the kind of salts they formed; by the shape of their crystals. Perhaps inspired by the card game Patience, he listed the elements and their properties on cards, arranging and rearranging them hundreds of times. By the beginning of 1869 he had found the key. At the March meeting of the Russian Chemical Society, Mendeleev made eight points, the first of which is his famous periodic law:

1. The elements, if arranged according to their atomic weights, exhibit an evident periodicity in their properties.
2. Elements with similar properties either have similar atomic weights (that is, they are neighbors within the same row of the periodic table), or have atomic weights that increase regularly (that is, they fall one above another in the same column of the table).
3. The columns of the table correspond with the valences, or "combining power" of the elements.
4. The elements most commonly found in nature all have low atomic weights and sharply defined properties. They head the table, and characterize the various families of elements.
5. The atomic weight of an element determines its character.
6. The discovery of many yet unknown elements may be expected.
7. The atomic weight of an element may sometimes be corrected by knowing its proper place in the table and the atomic weights of its neighbors.
8. Certain characteristic properties of the elements can be foretold from their atomic weights.

Dmitri Mendeleev.

Mendeleev, in contrast to Meyer and other chemists pursuing similar work, did not rest with generalizations. Instead, he fearlessly turned the periodic table into "an instrument of thought" that he used to make specific predictions. In building the table, Mendeleev was so sure of himself that when the properties of an element did not match those predicted by its atomic weight, he assumed that the weight, not his periodic law, was wrong. In order to fit indium into his chart, for example, he had to triple its accepted atomic weight. Similarly, he doubled the accepted atomic weight of uranium, and trimmed the accepted weight of titanium from 52 to just under 48.

Most chemists did not embrace Mendeleev's grand unifying vision. Many disliked his tampering with accepted atomic weights.

Meyer, for example, wrote in 1870, "it would be rash to change the accepted atomic weights on the basis of so uncertain a starting point." Many found his entire scheme unconvincing, even unscientific. But Mendeleev had thrown down a gauntlet that could not be ignored: "It was necessary to do one or the other," he wrote, "either to consider the Periodic Law as completely true and as forming a new instrument of chemical research, or to refute it."

The periodic law and table of the elements did win acceptance over the next two decades, largely because of Mendeleev's predictions. He put the work and his reputation on the line by predicting the atomic weights and detailed properties of three elements, eka-aluminium, ekaboron, and ekasilicon, to fill specific gaps in the table. When the French chemist Lecoq de Boisbaudran discovered gallium in 1875, its chemical properties matched Mendeleev's "eka-aluminium" beautifully, but it was lighter than predicted. Mendeleev insisted that the measurements must be wrong. De Boisbaudran repeated his experiments and found, to his surprise, that Mendeleev had foreseen the atomic weight of the new material more accurately than de Boisbaudran had first measured it. Scandium, Mendeleev's ekaboron, followed in 1879, and germanium, his ekasilicon, in 1886. In each case, his predictions proved accurate—not just atomic weights, but also the specific density and heat of the elements, the compounds they formed, and even the densities and melting points of those compounds. By 1889, when Mendeleev published the fifth edition of his textbook *Principles of Chemistry*, he was able to include a long list of experimentally verified predictions, and graciously laud the scientists whose work had validated his as "the true founders of the periodic law."

Mendeleev's periodic table gave chemistry three great gifts. First, he transformed the elements from a swarm of unrelated individuals into a well-ordered array in which both family traits and similarities to neighbors stood out clearly. Secondly, he stole from astronomers and physicists the power to predict. Like Leverrier, who had used Newton's laws to predict the existence and location of Neptune, Mendeleev used the regularities revealed by his table to predict the existence and properties of new elements. Nothing like that had ever been done. He risked ridicule, but won fame. That predictive power gave chemistry enormous confidence and creativity, leading to today's designer materials, designer drugs, and even designer genes. And once Mendeleev's work was accepted, it was clear that something of great importance, something fundamental, must be at work to marshal the elements into such neat ranks and files.

In 1870, neither Mendeleev nor anyone else knew why nature had created just this set of elements or had given them their particular properties. Mendeleev proved to be prophetic about the ultimate source of the periodicity he discovered. The key, he insisted, is that the elements vary discontinuously, without intermediates. He intuited that this unique stepwise arrangement of the elements and their properties must reflect a basic, still undiscovered law of nature. Concerning these unexplained leaps, Mendeleev wrote, "the time will come for their full explanation, and I do not think that it will come before the explanation of such a primary law of nature as the law of gravity."

As we now know, the new "primary law" Mendeleev foretold springs from the quantum world. In 1900, Max Planck took the first reluctant step into that realm by showing that energy can only be emitted or absorbed in discrete bits. In 1913, the Danish physicist Niels Bohr explained the position of elements in the periodic table by joining Planck's quantum of energy to Rutherford's model of the atom. In 1925, with the addition of the Pauli exclusion principle to the Bohr atom, physicists showed that four quantum variables, constrained to change in discrete steps, force electrons to accumulate in successive atomic "shells." For example, helium's two electrons form a self-sufficient shell, making it chemically inert. But lithium's third electron barely begins to fill the next shell, so it is highly reactive. With increasing atomic number (and weight), the shells fill up in runs of two, eight, eighteen, and thirty-two steps. Quantum rules place the elements one by one into the rows and columns so neatly laid out by Mendeleev fifty-five years earlier.

Mendeleev's strength of character was as clear in his politics as it was in his bold scientific predictions. Growing up in Siberia, he had absorbed the idealism of the radicals and reformers exiled there. His sister had married an exiled Decembrist, one of those who had rebelled against Tsar Nicholas I in December 1825. As Mendeleev's scientific reputation grew, his lifelong advocacy of humanitarian and democratic ideas became more threatening to the academic establishment. In 1880, reactionaries within the St. Petersburg Academy of Sciences caused an uproar by blocking his election to full membership. In 1890, Mendeleev took the extraordinary step of personally delivering a petition from the students at the University of St. Petersburg to the Ministry of Education. Despite his international repute and thirty-year career at the university, the authorities spurned his intervention, and he was forced to resign.

Mendeleev never distanced himself from his roots. Throughout his life he continued to dress in simple, coarse clothing and disdain the trappings of the elite. After meeting Mendeleev, the British chemist William Ramsay, who discovered the noble gas group of elements, described him as "a peculiar, hairy foreigner." Mendeleev poured decades of work into studies of Russia's undeveloped oil, coal, and iron resources. He believed that economic development was the only way to pry the Russian people from the jaws of poverty. After years of indifference on the part of the Tsarist government, Mendeleev's work eventually stimulated Russia's early industrial development. Like his contemporaries, Tolstoy and Dostoyevsky, he spoke and wrote in support of human rights. But he was always a pragmatist, stressing education, employment, and economic development as the keys to a better world. And he believed passionately in science. It was in the laboratory, pursuing scientific truth, that he forged his own character and destiny. "Knowing how contented, joyous, and free is life in the realm of science," he wrote, "one fervently wishes that many would enter its portals."

Mendeleev died of pneumonia in the winter of 1907, at the age of seventy-two. The university students to whom he had been so loyal marched by the hundreds in his funeral procession, proudly carrying his coffin and the periodic table. It was a unique moment in the history of Russia and of science. Still, judging by the way he had lived, Mendeleev would not have been impressed. "Fine," he might have said. "Now back to work."

20

In the Realm of Radioactivity

But I, even I, keep a sort of hope
that I shall not disappear into nothingness.

—Marie Curie, age 22

Sometime in 1885, eighteen-year-old Marya Salomée Sklodowska and her older sister Bronia hatched a desperate plan. Marya, a brilliant young woman who had gone as far as she could within Poland's educational system, would work as a governess to allow Bronia to study medicine in Paris. As soon she could, Bronia would help Marya come to Paris to study science. As young women from an impoverished family in Russian-occupied Poland, this was their only hope of realizing their potential as anything besides wives and mothers. For Marya, the next five years were ones of mind-numbing work as a governess under emotionally trying circumstances. She almost gave in to the forces conspiring to turn her into "a nullity." But she clung to what she called her first principle—"Never to let one's self be beaten down by persons or events." She was twenty-three when she at last climbed into the fourth-class ladies' carriage for the three-day journey from Warsaw to Paris. There she would change her name to Marie, earn top honors in physics and mathematics at the Sorbonne, marry and lose the love of her life, endure a shattering scandal, and perform the groundbreaking studies that would win her two Nobel Prizes.

Marie Curie (1867–1934) had married fellow scientist Pierre Curie and given birth to their first child before she started her doctoral research. She chose to investigate an obscure finding, the mysterious rays that Henri Becquerel had discovered in 1896. He had stumbled on them when he found that uranium could fog a photographic plate through thick black paper. Becquerel thought the rays were a new kind of phosphorescence—a delayed emission of energy absorbed from another source. In England, the seventy-three-year-old Lord Kelvin found that the rays from uranium, like X-rays, "electrified" air as they passed through. That's all that was known until the winter of 1897, when Marie picked up the trail.

Marie had the advantage of an instrument Pierre had invented that allowed her to make precise measurements of minute electric currents. With that, she set out to measure the ionization caused by the emanations from uranium and other substances. From the start, her work was precise, systematic, and insightful. She quickly found that most elements did not emit these rays—as she put it, they were not "radioactive." One element, thorium, emitted more radioactivity than uranium. And, surprisingly, so did pitchblende, a uranium-bearing ore with a black, pitch-like luster. That, she realized, meant that pitchblende might contain new, highly radioactive elements.

It was clear that Marie was exploring important new territory. Pierre set aside his work on crystals to join her—the start of years of highly productive collaboration. On Tuesday, April 12, 1898, the French Academy of Sciences heard Marie's first paper on radioactivity. Since neither Pierre nor Marie belonged to the Academy, Gabriel Lippmann, Marie's teacher and an Academy member, read the paper. The groundbreaking report not only suggested the existence of one or more new elements, it was the first to use radioactivity as a tool for their discovery. Most importantly, having found that the amount of radioactivity depended only on how much radioactive matter was present, and was not affected by chemical reactions, Marie was the first to realize that radioactivity must be a fundamental atomic property, not a chemical one.

With her typical determination, Marie set out to prove the existence of the new element or elements. That meant extracting and purifying enough of the materials to detect their spectral signatures, then isolating still more to determine their atomic weights. The method she developed was to dissolve pitchblende to produce a solution of chloride salts. When the solution crystallized, the radioactive atoms appeared within chlorides of bismuth and barium. She assumed that the radioactive elements would crystallize at slightly different rates than the compounds in which they were hiding. So she repeatedly dissolved and recrystallized the solutions. Over time, and with great effort, she was able to extract minute quantities of two new, intensely radioactive elements. From the bismuth crystals came polonium, and from barium emerged radium. The elements revealed themselves in their unique spectral lines and by their radioactivity—hundreds of times more intense than that of uranium. But it became clear that nearly ten tons of pitchblende would have to be processed to produce measurable amounts of either element.

Marie Curie took that task on herself. It meant three years of exhausting labor in an unheated warehouse, stirring huge vats of boiling chemicals with a heavy iron paddle—then painstakingly crystallizing and re-crystallizing the solutions. "I would be broken with fatigue at the day's end," she wrote. But the results were spectacular. By July 1902, she had isolated one-tenth of a gram of radium. She measured its atomic weight as 225.45 (it's now known to be slightly heavier, 226), and could place it firmly within Mendeleev's periodic table as one of the alkali earths. She and Pierre would sometimes go by the lab at night to admire their handiwork. "Our precious products," Marie wrote, ". . . were arranged on tables and boards; from all sides we could see their slightly luminous silhouettes, and these gleamings, which seemed suspended in the darkness, stirred us with ever new emotion and enchantment."

The Curies, along with other researchers, recognized that the big question was where that gleaming energy came from. As early as July 1900, Marie had written that radioactive materials might be "substances in the course of breaking up." She tried, unsuccessfully, to measure the weight loss that would have to accompany the release of energy. Early in 1903, Pierre was the first to measure the "extraordinary" amount of heat radium continually emitted. Still, as bold as Marie was as an experimenter, she was cautious about theorizing.

It was Ernest Rutherford, working at McGill University in Montreal, Canada, who in 1903 convincingly explained radioactivity as a product of the transformation of one element into another. With many scientists by then racing to study radioactivity, it was inevitable that discoveries would also be made by others. But Marie Curie always felt a proprietary interest in radioactivity, and kept her place in the forefront of the field until her death. One sign of this was her impact on the scientific vocabulary. She coined the terms "radioactive," "radioactivity," "disintegration," and "transmutation" to describe these new phenomena.

Marie Curie.

The year 1903 was a time of scientific triumph for the Curies. In May, Marie defended her dissertation at the Sorbonne. She was able to point out that "our researches have given rise to a scientific movement." In November, she and Pierre received the British Royal Society's prestigious Sir Humphry Davy medal. And, within weeks, they learned that they had won the Nobel Prize, along with Becquerel, for "their joint researches on the radiation phenomena." Marie became the first woman to receive the Nobel Prize. (Thirty-two years later, her older daughter, Irene, would become the second female laureate in the sciences.) But ominously, Marie was too sick to travel to Sweden; her work in the lab was taking its toll.

The Nobel Prize proved a mixed blessing. The Curies were able to use the award money to help finance their research, and their new-found status helped them to advance the cause of science within France. The French press seized upon the prize as a patriotic victory, and transformed the intensely private Curies into public heroes. Marie had suffered a miscarriage a few years earlier, but carried their second child, Eve, to full term. Like working mothers today, Marie had to make painful choices among competing responsibilities. It must have been torture for her and Pierre to read pundits debating which of them had done the important scientific work and arguing about whether she could be both a scientist and a mother. They bitterly resented the intrusion of the press and public into what had been their own private world.

That world was shattered on April 19, 1906, when Pierre was struck and instantly killed by a heavy, horse-drawn wagon. "I lost my beloved Pierre," Marie wrote twenty-three years later, "and with him all hope and all support for the rest of my life."

The face Curie presented to the world was a formal one—always the scientist, never the person. That may have been necessary, for the world was not ready to accept her as both a scientist and a woman. Her self-imposed mask finally fell away in 1990, when the journal she kept during the year following Pierre's death was finally made public. Only in its pages, in private, could she reveal herself. "He is gone forever," she writes a few weeks after the accident, "leaving me nothing but desolation and despair." She rejects the idea of suicide, but yearns for death; "Among all these carriages, isn't there one which will make me share the fate of my beloved?" As the months pass, she gradually resumes her duties as mother, teacher, and researcher, but with the aching absence of joy of someone mortally wounded. "Never will I have enough tears for this," she writes. A full year after his death, she is still grieving. "The grief is mute but still there. The burden is heavy on my shoulders. How sweet it would be to go to sleep and not wake up. How young my dear ones are. How tired I feel!"

As a prominent woman scientist, Curie could not escape the sexism of her day. She incurred an enormous dose in 1911 when she was nominated to the French Academy of Sciences. The press treated her candidacy as if it would destroy the traditional roles of women as self-effacing mothers, seductive sex objects, or inspirational muses for men. The right-wing press, having tasted blood in the infamous Dreyfus affair fifteen years earlier, cast her as the dupe of a bizarre Jewish-Huguenot

conspiracy against French Catholicism, represented by the other leading candidate, an elderly physicist, Eduoard Branly. Even after the
Academy rejected Curie, the press continued to scold her. She had
failed to display the reserve and self-effacement proper to a woman,
one editor railed. She had gotten what she deserved, a "lesson in patience and modesty."

The next "lesson" she received nearly killed her. Late in 1911, the
news broke that she had been having an affair with a married man,
fellow physicist and longtime friend, Paul Langevin. We now know
that their relationship started years after Pierre's death, and involved a
deep connection that brought a few moments of joy into the lives of a
lonely widow and a man trapped in a terrible marriage. But, at the
time most of the information came from Langevin's wife and his
mother-in-law, who depicted Curie as a heartless home wrecker. Further humiliation followed when stolen letters between the lovers sur-
faced. The right-wing French press went wild, turning their private
affair into an attack on marriage, motherhood, and France itself. It
showed, one editor wrote, "France in the grip of the bunch of dirty for-
eigners, who pillage it, soil it, and dishonor it." At one point, an angry
mob besieged her home.

In the midst of this turmoil, Marie learned that she had won a second Nobel Prize. But days later Svante Arrhenius, representing the
Nobel Academy, advised her not to come to Sweden or accept the
award until she had cleared her name. Courageously, Curie attended
the ceremony and accepted the award in person. "I believe there is no
connection between my scientific work and the facts of private life," she
wrote. In her acceptance speech she gave credit to others where it was
due, but for the first time unashamedly took credit for her own accomplishments. A decade of intense research, she said, had "furnished proof
of the hypothesis made by me, according to which radioactivity is an
atomic property of matter and can provide a method for finding new elements." And, poetically, "Radioactivity is a very young science. It is an
infant that I saw being born, and I have contributed to raising with all
my strength. The child has grown; it has become beautiful."

On her return from Sweden, Curie collapsed. She had developed
a severe kidney infection and needed surgery. It took her two years to
return to her research and teaching. Still, during World War I, she was
well enough to organize and implement a mobile radiological service
that eventually provided X-rays to more than a million injured front-
line soldiers. Her teenage daughter Irene worked with her, starting a

collaboration that would last until Marie's death. Curie's remarkable war work redeemed her reputation, except among the most rabid French xenophobes. After the war, Marie demonstrated her organizational and managerial skills as head of the new Institute of Radium in Paris. She developed it into a leading research center and trained a new generation of scientists, many of them women. It was there that Irene and her husband, Frederic Joliot, became the first researchers to transmute one element into another artificially. Curie became the first woman to teach at the Sorbonne, and the first woman elected to the French Academy (although in medicine, not in science). She twice traveled to America, where she was hailed as "the greatest woman in the world." At the same time, an American correspondent described her as having "the saddest face I had ever looked upon."

The years, the traumas, and her lifelong exposure to radiation took their toll. She left her laboratory for the last time late in May 1934. Her doctors diagnosed aplastic anemia. Two months later, at the age of sixty-six, she was dead. In her later years, few people saw beyond her public mask. But Einstein, never one to be fooled by appearances, wrote to her that he had found her "full of goodness and obstinacy, and it is for that I like you, and am happy to have been able . . . to glimpse the depth of your mind where everything gets figured out in private."

Like her precious radium gleaming in the dark, Curie's discoveries continue to shine. They stand as a remarkable accomplishment for an impoverished governess who in her youth hoped against hope that she would not disappear into nothingness, and, in the dark years after losing her beloved Pierre, often wished that she could.

21

Planck's Quantum Leap

[T]he whole procedure was an act of despair
because a theoretical interpretation had to be found at any price,
no matter how high that might be.

—*Max Planck, 1901*

Max Planck (1858–1947) did not set out to start a revolution. The descendant of a long line of conservative, high-minded churchmen and scholars, he told one of his professors that he would be glad simply to understand and perhaps deepen the existing foundations of physics. But at the age of forty-two, in the midst of his career, he found that he could not explain a seemingly simple problem unless he took the radical step of chopping energy up into tiny packets, or quanta. Soon these minute units showed up wherever physicists looked, and convincingly explained all kinds of things that classical physics could not. The packet, the quantum, that appeared in Planck's equations, proved to be a new universal constant, as fundamental as the speed of light. It propelled physicists into a strange new world—a world in which light is both a particle and a wave, in which a measurement here and now can instantaneously determine an event far away in space and time, in which space constantly churns with evanescent particles, and in which, perhaps, an infinite number of universes are born every instant.

Rather than deepening the foundations of classical physics, Planck's discovery forced them to be abandoned and rebuilt from

scratch on the shimmering sands of quantum uncertainty. The resulting edifice—almost all of modern physics and technology—is more
powerful and brilliant than ever, but vastly more mysterious.

Like Einstein, Planck homed in unerringly on the foundations of
physics. The beacons that guided him were great regularities such as
the first and second laws of thermodynamics, which he trusted as absolute truths. In the waning years of the nineteenth century, Planck
suspected that something absolute lurked in what was called blackbody
or cavity radiation—the light and heat radiating through a small opening in a heated container. The intensity of the radiation depended only
on the wavelength being measured and the temperature of the container; the shape, size, or material of the container did not matter.
Planck was guided by the same kind of intuition that had led Pythagoras 2,400 years earlier to see something absolute, which he called number, in vibrating strings. (In both cases, matter seemed to be nothing
more than the visible expression of something more fundamental.)

Blackbody radiation was one of the nagging loose ends of classical
physics. The applicable theory predicted that all the energy in the container would be concentrated at the shortest, most energetic wavelength—the so-called ultraviolet catastrophe. If the theory were right,
opening your oven door would zap you with a fatal dose of high-energy
radiation. Since that did not occur, it was clear that nature was able to
spread the energy out across the spectrum. But nobody knew how.

Planck worked on the problem for five years starting in 1895. By
October 1900, he had a partial solution. He had found a way to cob-
ble together two formulas, one of which matched the intensity curve
for longer wavelengths, the other for shorter wavelengths. The new
formula did a great job matching what experimenters had found, but
Planck was far from satisfied. He spent the next two months on a desperate quest to derive the formula from basic physical assumptions.

He succeeded, but at a high cost. In order to derive the correct radiation law, Planck had to jettison two fundamental assumptions. First
he had to reject the second law of thermodynamics as an absolute
truth and replace it with a mere probability—the ability of a system to
do work does not *always* decrease over time, it *almost* always de-
creases. Then he was forced to make an even more radical break. He
had to treat energy—something that had always been assumed to vary
smoothly and continuously—as if it came in tiny packets, which he
called quanta. He took the same step for energy that Leucippus had
taken 1,300 years earlier for matter. It was as if energy no longer

Max Planck.

flowed like a river; instead, nature delivered it only in cans. That concept proved to be the Trojan horse that would eventually topple the
towers of classical physics.

Planck presented his derivation of the quantum of action in a lecture to the Berlin Physical Society on December 14, 1900. At that time
he went no further than to assume that when light and matter interact
energy can only be absorbed or emitted in multiples of his fundamental unit. Driven by the logic of his equations, Planck had taken the first
step into the quantum realm. Although he knew that he had made a
radical assumption, neither he nor anyone else knew where it would
lead. For the most part, it fell to others to discover those implications.

Einstein was the first to pick up the quantum baton. In 1905, he found that Planck's quanta neatly explained the photoelectric effect, in which light falling on a metal surface can create a flow of electricity. Einstein boldly applied quantum theory not just to the interaction between matter and radiation, but to light itself. In 1909, he declared that light consisted not only of waves, but simultaneously of a stream of discrete particles. It was those particles, he found, that packed enough punch to separate electrons from atoms to produce photoelectricity. Einstein viewed this, not relativity, as his only truly revolutionary work. For many years most physicists, including Planck, believed that Einstein had gone too far—the wave theory of light was too successful to be challenged. It was only in 1922, when the American physicist Arthur Compton showed experimentally that X-rays could behave like billiard balls, that physicists realized that light particles— photons—were just as real as electromagnetic waves. Ironically, the conservative Planck never fully accepted the reality of photons.

The Danish physicist Niels Bohr made the next great advance. In 1913 he described the hydrogen atom in quantum terms. By assuming that electrons could only carry certain amounts of angular momentum, integral multiples of Planck's constant, he did what classical physics could not—he showed how an atom could be stable. Maxwell's equations described most electromagnetic phenomena beautifully, but they predicted that electrons would radiate away all their energy and crash into the nucleus in a tiny fraction of a second, which of course doesn't happen. Planck made one additional assumption, that electrons emit light or other radiation only when they jump from one energy level to another—the quantum leap. This allowed him to explain the exact spectrum of the light given off by excited hydrogen atoms. His predictions matched measurements to one part in a thousand.

It took a quarter century before quantum theory truly came into its own. In 1925 and 1926 physicists developed two very different mathematical approaches to the quantum realm. Stimulated by the earlier work of Einstein and Louis de Broglie, Erwin Schrödinger took seriously the idea that particles are nothing but waves. From that, he produced his wave equation, which described quantum events using the same kinds of equations physicists use to analyze any kind of wave. The equations were familiar to physicists, but not what they described. Schrödinger's waves represented the *probability* that a particle would be detected at a certain location—the quantum uncertainty that Einstein refused to accept. Still, the approach proved to be immensely powerful, and led to much of modern physics.

At the same time, Werner Heisenberg found that a completely different kind of mathematics—matrix algebra—also appeared to underlie the quantum world. Using matrices, Heisenberg, Pascual Jordan, Max Born, and Paul Dirac developed a systematic way of understanding and treating quantum events that was just as powerful as Schrödinger's wave equations. Dirac, among others, soon proved that the two mathematical systems were in fact equivalent; they contained the same basic assumptions and produced identical results. Together, they provided physicists with a formidable set of tools for deciphering the world of the very small.

However, physicists were quick to point out that neither approach made the quantum world any less mysterious. Subatomic particles, atoms, and (we know now) even molecules behave exactly like waves in certain circumstances and exactly like particles under other circumstances. Niels Bohr's concept of complementarity describes the phenomenon: objects act like particles or like waves depending on how we detect or measure them. But neither description tells the whole story. Waves and particles seem to be two faces of a shape-changing beast whose "real" nature we may never know, but whose behavior follows quantum rules.

The quantum proved to be Planck's greatest scientific achievement. After 1900, he devoted an increasing amount of his time and thought to administrative and leadership roles in science. In 1918, he won the Nobel Prize for his discovery. He headed the mathematical and physical sciences section of the Prussian Academy of Sciences for over three decades, and also served as president of the Kaiser Wilhelm Gesellschaft, the center of German science, for many years before World War II, and again, just after the war. Planck felt that it was his duty to try to protect and preserve science in Germany in the face of attacks by those who were marching to the Nazi drumbeat. He was courageous enough to meet with Hitler to try to stop him from driving Jewish scientists out of Germany, a confrontation which provoked one of Hitler's infamous tirades. Planck paid a high personal price for his decision to remain in Germany. His home was completely destroyed in a bombing raid. His last surviving child, Erwin Planck, was executed in 1944 because he was suspected of involvement in a plot to assassinate Hitler. And, as chaos overtook Germany, Planck and his wife were reduced to hiding in the forest and sleeping in haystacks. He was then nearly ninety years old.

Today, the quantum mechanics that flowed from Planck's "desperate" step serves as the foundation for our understanding of the world. The theory tells us how the molecules, atoms, and subatomic particles of which everything is composed behave. It also describes the behavior of some full-sized systems, such as superconductors or globs of ultra-cold atoms called Bose-Einstein condensates. These equations, which depict reality as mere probability, have given scientists the ability to tap the energy of atoms, to understand and predict chemical reactions, and to build electron microscopes, computers, lasers—in fact, much of our current technology. Economists estimate that thirty percent of the U.S. gross domestic product is based on quantum technologies. Today, scientists are on the verge of building quantum computers that will use the ability of quantum systems to be in many states at once to solve currently intractable problems. And at some point, perhaps in this century, physicists tell us that quantum theory and general relativity will be smelted into a single theory capable of explaining both the very small and the very large, and of tracing the history of the cosmos all the way back to its first instant—the so-called theory of everything. If we find the quantum world strange, the reality revealed by the theory of everything promises to be stranger still.

22

Wired on
Wireless

When I am allowed to work in my own technical field
I am the happiest man in the world.

—*Guglielmo Marconi, 1926*

Except in his mother's eyes, Guglielmo Marconi's future greatness was invisible. Marconi (1874–1937) was an indifferent student, preferring his own dreams to his tutor's lessons. His woolgathering and willfulness infuriated his Italian father, but his doting Irish mother saw them as earmarks of genius. To his father's disgust, he failed to qualify for the University of Bologna or the Italian Naval Academy. But in young Marconi's case, his mother's intuition and determination ruled. They seem to have given him a faith in his own vision that allowed him to overcome his failures and vault seemingly insurmountable barriers throughout his life.

Marconi was twenty when his first great insight struck him. While on vacation in the Italian Alps, he read an article about Heinrich Hertz, a brilliant physicist who had died in January 1894 at the age of thirty-six. Hertz had found a way to generate and detect the electromagnetic waves that had been predicted by Michael Faraday and described mathematically by James Clerk Maxwell. Hertz noticed that when a high-voltage spark crackled across what he called an exciter, a faint spark flickered across a nearby detector, two closely spaced metal balls joined by a loop of wire. Hertz had made Maxwell's invisible waves real, and given scientists a tool with which to study them.

The spark that flashed through Marconi's mind that day, however, was a way to use the mysterious waves. He envisioned putting them to work to carry information through the air at the speed of light—wireless communication. From that moment on he was driven to make wireless telegraphy a reality. In single-mindedly chasing that goal he made or stimulated most of the discoveries and advances that led to radio, television, radar—in fact, to the intricately connected wireless world we live in today.

When Marconi returned home at the end of that summer, he commandeered the attic of his parents' house as his laboratory. He set out to reproduce what scientists had already accomplished, and soon was able to make a bell ring by detecting electromagnetic waves from a spark thirty feet away. True to form, the first person he demonstrated this to was his mother. Within months he pushed the range of his exciter and "coherer"—an improvement on the detector used by Hertz—to a few hundred yards. At that point, he won some grudging financial help from his father, and was able to enlarge and refine his equipment.

Marconi soon began to make discoveries of his own. He improved the coherer by tinkering with its materials, shrinking it, and creating a vacuum inside it. He increased the range of his transmitter a bit by adding metal plates to each side of the spark gap. By accident, he held one plate high in the air while resting the other on the ground. Suddenly his signals carried across the fields for close to a mile. With this new elevated aerial and ground, he soon pushed his signal more than a mile, and even over a hill. By the fall of 1895, at the age of twenty-one, he had turned his flash of inspiration into a working model.

Marconi next approached the Italian government, hoping to gain its support in order to turn his idea into a usable communication system. Through his father's contacts he contacted the Ministry of Posts and Telegraphs, only to be told that they had no interest in the invention. Marconi was bitterly disappointed, but, characteristically, not discouraged. Through relatives of his mother, he learned that the English might be more receptive.

In February 1896 Marconi and his mother traveled to London. His mother's nephew, Henry Jameson-Davis, put Marconi in touch with several people who were in a position to help him. By June, Marconi had rebuilt his apparatus, which had been smashed by suspicious customs inspectors, had written a painstaking patent application, and had won a provisional patent—the first anywhere for wireless telegraphy. Until then, he had lived in fear that one of the established scientists studying electromagnetism would beat him to his goal. The scientists of his day, however, were more interested in deepening their understanding of electromagnetism than in applying it. One of them, Oliver Lodge, later wrote, "whereas we had been satisfied with the knowledge that it could be done, Mr. Marconi went on enthusiastically and persistently . . . till he made it a practical success."

Luckily for Marconi, William Preece, a remarkably open-minded and farsighted man, was engineer-in-chief of the British Post Office. He met the young man, studied his apparatus carefully, and arranged for a public demonstration. On July 27, 1896, Marconi set up his equipment on the roof of the General Post Office and atop another Post Office building nearly a mile away. In front of Preece and a number of engineers and officials, Marconi successfully sent and received a series of messages. "Young man," Preece said, "you have done something truly exceptional. I congratulate you on it."

Now backed by the Post Office, the twenty-two-year-old Marconi pushed even harder to increase the power and range of his system. By September of that year he was sending messages miles across Salisbury Plain. Both Preece and Marconi believed in the potential for wireless communication to save lives at sea, and quickly turned to tests over water. By May 1897, Marconi's signals spanned nine miles between the mainland and the island of Flatholm in the Bristol Channel.

In the next few years, Marconi formed the company that would later be known as Marconi's Wireless Telegraph Company, Ltd. He increased the range of wireless communication through experiments in England and Italy—which had belatedly become interested in him. His company began turning out wireless equipment from a factory in Chelmsford, England, and built the first of what would become a worldwide network of wireless stations. He also made a major technical advance that led to his famous patent no. 7777, for tuned wireless telegraphy. That allowed him to control the frequency of the electromagnetic waves he was sending.

From the start, Marconi had an uncanny instinct for publicity. He gained wide notice by providing the first wireless coverage of a sporting event—the Kingston Regatta in July 1898. He scored another coup by putting Queen Victoria in touch with her son, Edward, Prince of Wales, who was recuperating from an injury aboard the royal yacht. Wireless demonstrated its lifesaving capabilities for the first time when the newly installed station aboard the East Goodwin Lightship, a floating lighthouse, saved the crew of a grounded ship.

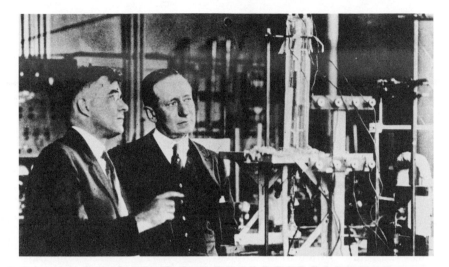

Guglielmo Marconi (right) and Irving Langmuir (left).

And in March 1899 he sent messages across the English Channel.

Marconi's investors were eager to develop Marconi's technology into a money-making business, but Marconi had more grandiose plans. He was determined to be the first to send wireless signals across the Atlantic. Most scientists believed this was impossible. They expected electromagnetic waves, like light, to be blocked by the curvature of the earth. As usual, Marconi's determination overcame all opposition. Driven by the inventor, the company sank a huge amount of its capital into building two enormous, high-power stations, one at Poldhu on the Cornwall coast, the other on Cape Cod.

Nature, however, intervened. In September 1901, a powerful storm swept over Poldhu, smashing the ring of 200-foot-tall aerials Marconi had built. In November, the same fate struck the sister complex on Cape Cod. Undaunted, Marconi reassured his business partners, designed and built a simpler aerial at Poldhu, and sailed to Newfoundland, Canada, where he hoped to receive signals with the simplest of equipment, a strand of wire dangling from a balloon or a kite. He landed on Friday, December 6, 1901, and by Wednesday was ready for the first trials. Again the weather worked against him, ripping the balloon from its moorings. The same thing happened to the first kite they launched. But on Thursday, December 12, 1901, at

12:30 P.M., Marconi and his assistant, George Kemp, managed to keep a second kite and aerial aloft long enough to hear the prearranged signal, "dit-dit-dit," the letter S in Morse code, from across the Atlantic.

Marconi had been famous before this, but he now became a kind of scientific superstar—the wireless wizard. He achieved the kind of adulation that would lionize Charles Lindbergh twenty-six years later. Marconi's fame soared even higher in 1912, following the sinking of the *Titanic*. The tragedy would have claimed many more lives if his equipment had not been on board to send out a call for help.

A bit of Marconi's genius can be seen in a remarkable about-face he led his company through in the 1920s. Until then, the thrust of his work had been to design and build more and more powerful transmitters capable of driving long-wave signals thousands of miles. But in the 1920s, he returned to the short-wave signals of his earliest experiments. He quickly found that they could be concentrated into a directed beam rather than broadcast, and, although no one knew how, carried well beyond the horizon. At the same moment that his company was building enormous long-wave stations around the world, Marconi took the huge risk of switching to a largely untried short-wave system. He quickly negotiated a test of the system for the British government. In one of his now-familiar crash programs, Marconi pushed his engineers to overcome a swarm of technical obstacles. By October 1926 he had established efficient and reliable two-way communication across the Atlantic using short wavelengths. Once again, Marconi had created an entirely new technology.

Marconi would be instantly accepted in today's Silicon Valley. His visionary pursuit of the next breakthrough, his high-risk venture capitalism, and his crash programs to develop and implement his technological dreams scandalized Victorian gentlemen-scientists. But, like wireless communication itself, they have come to define today's world.

It's not surprising that Marconi's amazing technical and professional successes were purchased at the expense of his personal life. He had an amazing capacity to focus single-mindedly on his research, and an equally remarkable ability to tune out the people closest to him. Those included his first wife, Beatrice O'Brien, and the three children from that marriage. His politics, too, appear less than savory from today's perspective. He was a patriotic Italian, and effortlessly transferred his allegiance to Mussolini and the Fascists who came to power in 1922. Although Marconi died before World War II and the Holocaust, he was an outspoken supporter of Mussolini's infamous war on Abyssinia.

According to his doctor, Marconi was a scientist to the end. On his deathbed, he refused to accept the doctor's hollow reassurance that the absence of a pulse in his arm was due only to its position. "No, my dear doctor," he said, "this would be correct for the veins but not for an artery." He died moments later, at 3:45 A.M., July 20, 1937. At his funeral, as had been the case on many occasions during his lifetime, crowds surged through the street to be near him, and public figures lavished praise on him. But it was the British, first to recognize him, who gave him his most fitting farewell. On July 22, at 6:00 P.M., radio communication throughout the British Empire fell silent. For two minutes the airwaves were quiet, just as they had been before Marconi changed the world forever.

23

Rutherford Dissects the Atom

Rutherford had no cleverness—just greatness.

—*James Chadwick*

I am a simple man and I want a simple answer.

—*Ernest Rutherford*

In 1902, the thirty-one-year-old Ernest Rutherford (1871–1937) wrote to his mother: "I have to keep going, as there are always other people on my track. I have to publish my present work as rapidly as possible in order to keep in the race." He had some formidable competitors, including Marie and Pierre Curie. The race was to understand radioactivity and what it said about the nature of matter and energy. The winner was clear. In the course of the next thirty-five years, Rutherford would head three physics laboratories, building each into a humming center of research. Through a series of brilliant experiments, he would be the first to explain radioactivity as the product of atomic disintegration, the first to see the atom as a massive nucleus surrounded by electric charge, the first to identify alpha and beta particles, and the first to infer the existence of the neutron. Rutherford was the pioneer who opened the atom and its nucleus to human understanding.

Rutherford was born in rural New Zealand, the fourth of twelve children. His mother taught school, and his father farmed, ran a mill,

Ernest Rutherford.

and made rope. Like the young Newton, Rutherford was handy—he liked to take things apart and build working models of machines. He excelled at school and won several scholarships. One of his teachers, Alexander Bickerton, captured him perfectly. "Mr. Rutherford has a great fertility of resource," he wrote. "Personally, Mr. Rutherford is of so kindly a disposition and so willing to help other students . . . that he has endeared himself to all who have been brought in contact with him." Bickerton's glowing recommendation helped win Rutherford a scholarship to be the first "research student" at Cambridge.

At Cambridge, Rutherford made a splash by demonstrating, two years before Marconi, a device to detect Hertzian waves. Rutherford had left his sweetheart, Mary Newton, in New Zealand, and wanted to be able to marry her. He hoped to make some money with his invention. But, when J. J. Thompson, the head of the Cavendish Laboratory, tapped him for a study of X-rays, Rutherford set his radio work aside. By 1898, when he left Cambridge to head the physics laboratory at Montreal's McGill University, Rutherford had focused on the mysterious uranium rays Becquerel had discovered, and which the Curies were avidly investigating. Rutherford's fiancée had to wait until he fetched her from New Zealand in the summer of 1900.

Rutherford was a natural experimenter. Faced with a question, he immediately saw how to shed light on it. By placing different amounts of material between a source of radiation and a simple detector, he quickly distinguished two different kinds of radiation that he named alpha and beta rays. Alpha rays were easily stopped; beta rays were far more penetrating. At McGill, working with Frederick Soddy, a gifted chemist, Rutherford sorted out the rapidly growing collection of radioactive substances. He found that uranium, thorium, and radium, plus the radioactive gases they gave off, were way stations on evolutionary paths followed by radioactive elements. Each radioactive substance, they found, evolved at a characteristic speed and passed through a specific series of transformations on its way to a stable form. When an atom spat out an alpha particle, losing two units of positive charge, or a beta particle, losing one unit of negative charge, it changed from one element to another in predictable ways.

In 1902, Rutherford and Soddy published their results in two papers boldly entitled "The Cause and Nature of Radiation." From the fifth century B.C., scientists had believed that atoms, if they in fact existed, were permanent and indestructible. Now Rutherford was asking them to believe that atoms behaved like criminals on the run, randomly shooting off projectiles and assuming completely new identities. The experimenters were overturning one of the foundations of physics. Some of their colleagues pleaded with them not to publish these wild ideas, because they might discredit the university. Surprisingly, Rutherford's transmutation theory quickly became the accepted view. As with all of Rutherford's work, it was simple, physically meaningful, and grew naturally from compellingly clear experiments.

In 1903, with what one of Rutherford's collaborators, Edward Andrade, calls "the foresight of genius," Rutherford recognized that an enormous amount of energy was present in all matter, and realized that the Sun must generate its energy from "processes of sub-atomic change." Clearly, Rutherford was thinking deeply about the implications of his discoveries. Another physicist, William Dampier, later wrote that Rutherford jokingly wondered if "some fool in a laboratory might blow up the universe unawares," by triggering "a wave of atomic disintegration." Toward the end of his life, Rutherford characterized talk of releasing atomic energy as "moonshine." But this may have been wishful thinking. In a speech delivered during the dark days of World War I, he expressed the hope that nobody would find out how to release the energy of the atom until "man was living at peace with his neighbors." It has turned out to be much easier to unleash the energy of the atom than to learn how to live in peace.

Rutherford was the first to realize that natural radioactivity could clear up a mystery concerning the age of the Earth. The illustrious Lord Kelvin had confounded Darwin by "proving" that the Earth would be much cooler than it actually is if it were more than 30 million years old. Rutherford, however, showed that heat from radioactive decay allowed the Earth to be far older. He first presented this revelation with Lord Kelvin in the audience, and dreaded Kelvin's reaction. He later told his biographer that when he mentioned the age of the Earth, "I saw the old bird sit up, open an eye, and cock a baleful glance at me! Then a sudden inspiration came, and I said Lord Kelvin had limited the age of the Earth, *provided no new source was discovered.* That prophetic utterance refers to what we are now considering tonight, radium! Behold! The old boy beamed upon me."

Rutherford left Canada to return to England in 1907. A year later, he won the Nobel Prize in chemistry for his work on radiation and transformation of the chemical elements.

It was at the University of Manchester, working with his gifted student Hans Geiger, that Rutherford made his next great breakthrough. By then, alpha particles had become his favorite tool for probing the atom. He had determined that they were relatively massive particles moving at roughly one-tenth the speed of light. Geiger had noticed that when alpha particles were fired through a thin metal foil, a few were deflected more than they should be. Without knowing what they might find, Rutherford assigned a young student, Ernest Marsden, to look for particles whose paths were drastically bent. Almost immediately Marsden found particles that were being turned by 90 degrees or more—and even some that seemed to be bouncing back from the foil target. Rutherford was shocked. "It was almost as incredible as if you fired a fifteen-inch shell at a piece of tissue paper and it came back and hit you," he said.

Rutherford thought about those rebounding particles for nearly two years. It was not until 1911 that he burst into Geiger's room, beaming, to announce that he now knew what the atom looked like. Rutherford's mentor, J. J. Thompson, had devised the most widely accepted picture of the atom, the "plum pudding" model. It consisted of a sphere of positive charge with electrons scattered inside it. However, such a fuzzy atom could not deflect alpha particles in the dramatic way Marsden had seen. Instead, Rutherford reasoned, all of an atom's positive charge must be packed into a very small central sphere, surrounded at a distance by an equal negative charge. Since most of the atom was empty space, the majority of alpha particles would sail straight through, as observed. But particles fired toward the center would be whipped around like comets skimming the sun. Using the same formulas that Newton had used to trace the paths of comets, Rutherford calculated how many particles should appear at different angles, including particles that "came back and hit you." Before the day was over, he had Geiger counting particles to check the theory. It worked perfectly. With his unerring physical intuition, Rutherford had discovered the atomic nucleus.

Rutherford's atom explained many things, but it had one glaring flaw. The laws of electromagnetism dictated that an orbiting electron would radiate away its energy almost instantly. Bohr, another of the great physicists drawn to collaborate with Rutherford, brilliantly solved that problem in 1913. Bohr applied Planck's quantum of energy to the atom. The rules of the quantum world, he showed, allowed electrons to survive, but only in specific orbits. The orbits were separated by fixed amounts of energy. Electromagnetic waves of a particular frequency were absorbed or released only when an electron jumped from one orbit to another. From a few basic assumptions, Bohr was able to calculate the frequency of each spectral line given off by the simplest atom, hydrogen, as its single electron jumped from one energy level to another. Harry Moseley, another of Rutherford's collaborators, linked each element's chemical properties to the arrangement of its electrons, governed by the charge of the nucleus. For the first time, by applying quantum rules to Rutherford's nuclear atom, scientists could explain both the electromagnetic and chemical properties of the elements.

The team Rutherford had attracted was pulled apart during World War I. Rutherford spent the war years working on ways to use sound to detect submarines. After the war, Rutherford replaced his old teacher, J. J. Thompson, as head of the Cavendish Laboratory at Cambridge. There, he made two other major advances: In 1920, he predicted the existence and properties of a third basic atomic particle, the neutron. (Yet another of his talented students, James Chadwick, would detect the neutron twelve years later.) And, by using alpha particles to bombard various gases, Rutherford found that normally stable atoms of nitrogen were emitting protons, or hydrogen nuclei. He realized that normal, nonradioactive atoms could absorb alpha particles, break apart in predictable ways, and change into other elements. In the course of a few years, Rutherford had discovered the nucleus and learned how to blast it apart. Amazingly, the apparatus Rutherford used to transmute nitrogen into oxygen could fit easily in his hands. That apparatus has evolved into the city-sized atom smashers scientists are using today to search for the fundamental building blocks of matter and recreate the conditions that existed in the first instant of creation.

Rutherford was still enthusiastically involved in science until his unexpected death in 1937, following an accident at home. Bohr, addressing a scientific congress in Italy, announced Rutherford's death with tears in his eyes. Rutherford's ashes were buried in Westminster Abbey close to Isaac Newton.

Throughout his life, Rutherford remained extraverted, optimistic, a natural leader. He was a big man with an even bigger voice. When he was in a good mood, he liked to march through the laboratory lustily singing "Onward, Christian Soldiers." He never lost his sense of being in a race and wanting to win. Delays infuriated him, and he would sometimes send people scurrying out of his office with an angry outburst. But his anger never lasted long, and was inevitably followed by sincere apologies. As one of Rutherford's many collaborators wrote, "It can truthfully be said of him, as of very few people, that he had no enemies."

24

Einstein: Matter, Energy, Space, and Time

I would spend weeks in a state of complete confusion; and
it was only with great difficulty that I overcame the stupor
provoked by my first encounter with such questions.

—Albert Einstein

There is no logical way to the discovery of these elemental laws.
There is only the way of intuition, which is helped by a feeling
for the order lying behind the appearance.

—Albert Einstein

It was a sailor's compass, a gift from his father, that first aroused Einstein's curiosity about nature. Even at age four, the invisible force guiding the magnetized needle made an indelible impression on him. "There had to be something behind the objects, something that was hidden," he recalled. Within two decades Einstein (1879–1955) would penetrate nature's veil of appearances to reveal some of its deepest secrets—the elastic, interwoven fabric of space and time, and the undreamed-of equivalence of matter and energy.

A comforting myth depicts Einstein as a slow learner. That's hardly the case, although it is true that he didn't speak until he was three. He had an extremely keen mind from childhood. But he also disliked and distrusted authority and hated the rote learning that dominated his school years. Most of his teachers did not appreciate his stubborn independence of thought, the most notorious being a Dr. Joseph Degenhart, who found that Einstein's mere presence undermined his authority, and told Einstein that he would never get anywhere in life. Seldom has a prophecy proved so wrong.

Einstein experienced a period of deep religiousness as a child, quite at odds with the feelings of his assimilated Jewish parents. But by the time he was twelve, popular science books convinced him that the stories in the Bible could not be true. Still, he was inspired throughout his life by a profound awe at the underlying order of nature—he called it his Cosmic Religion. (The young Einstein had a radically different reaction to a book of Euclidean geometry—he later referred to it as his "sacred little geometry book.")

Einstein's rejection of authority reached the level of rebellion by the time he was sixteen. His father's business failed, forcing the family to move from Ulm, in southwest Germany, to Pavia, Italy. Einstein was left behind to finish school. He resisted the school's oppressive discipline, and was seen as disruptive. On his own, he wangled a medical certificate letting him leave school, and made his way to Pavia. He also renounced his German citizenship. The determined teenager made a deal with his worried parents—he would study on his own, take the entrance exams for the prestigious Swiss Federal Polytechnical School in Zurich, and become a Swiss citizen.

Despite earning high scores in mathematics and science, Einstein failed his entrance exams the first time. He made up for his deficiencies within a year, entered the Polytechnical School, and, despite preferring to study on his own rather than attending classes, graduated—fourth in a class of five—in 1900. It took him two years to find a job, during which he berated himself for not being able to help his parents, who were having serious financial problems. He also became deeply involved with Mileva Marić, a troubled young fellow student whom he soon married, despite his mother's opposition. The father of a close friend from school, Marcel Grossmann, recommended Einstein to the director of the Swiss patent office in Bern. So, in June 1902, greatly relieved, he took the job of technical expert, third class, on a trial basis. Luckily, his duties left him time to think.

His thoughts, or what the physicist Louis de Broglie described as Einstein's "inspired originality of mind," led him straight to the foundations of physics. The first results came in a flood of scientific papers published in the German *Annals of Physics* in 1905, a scientific "wonder year" comparable only to Newton's nearly 250 years earlier. Einstein, then twenty-six years old, produced four great papers, any one of which would have capped a physicist's career. Two of them were revolutionary.

First, Einstein built on Planck's hypothesis, published in 1900, that energy did not vary continuously, but instead came in minute, fixed packages. This concept turned out to be the foundation of what would become quantum theory, although at the time and for many years to come Planck viewed it as "a purely formal assumption." Einstein made those mathematical packets real, showing that treating light as a particle rather than a wave explained the photoelectric effect, one of the unsolved problems of nineteenth-century physics. A ray of light, he wrote, "consists of a finite number of energy quanta localized at points in space, moving without dividing and capable of being absorbed or generated only as entities." Unlike Planck, Einstein clearly recognized how revolutionary the idea was. It took two decades before other scientists reluctantly accepted the reality of the photon, and recognized that a new view of light as both wave and particle was a basic part of quantum reality. It was this work that earned Einstein the Nobel Prize in 1922.

The next loose end that Einstein took on was the luminiferous ether. The ether was a medium that scientists believed had to pervade all space in order to carry light and other electromagnetic radiation. It also served as a stand-in for Newton's absolute space, an immovable frame of reference within which motion took place, measured by the equally absolute flow of time. Newton's system was the bedrock of physics.

Einstein knew that a series of experiments had failed to find any sign of the ether. In 1887, the American physicists Albert Michelson and Edward Morley had carried out a definitive experiment using optical interferometry to try to detect the "aether wind" that was due to the Earth's motion. They failed to find it. But more importantly, Einstein also distrusted the idea on philosophical grounds. As early as 1899, he had concluded that it was a concept with no physical meaning.

Physicists had gnawed at this puzzle for more than two decades. They patched in expedient revisions, the most important being the Fitzgerald Contraction and the Lorentz Transformation. The Irish physicist George Francis Fitzgerald had shown that the ether wind would not be detected if the measuring apparatus, for unknown reasons, contracted in the direction of motion. Hendrick Antoon Lorentz, a Dutch theoretician, moved a step closer to relativity by showing how Maxwell's equations governing electromagnetic radiation could be transformed to work in moving systems.

Einstein boldly made these preliminary steps the basis of a new reality. He started his analysis by assuming that the laws of physics must remain the same in any system, at rest or moving, but not accelerating. A physicist in a rocket gliding through space should find that nature works exactly as it does in his laboratory on earth. This seemingly innocent assumption, he showed, led to earthshaking conclusions. First it forced the abandonment of absolute space, absolute time, and absolute motion, accepted since Newton. There simply was no special frame of reference on which to pin objects and events—all motion was relative. It made Fitzgerald's strange suggestion, that a moving object seems to shrink along its direction of motion, a reality. It made time as rubbery as space—time flows more slowly for objects in motion, and no two events can be said to be simultaneous. It showed that objects become more massive as they are accelerated. And it established a new absolute—nature's speed limit. Nothing can be accelerated faster than light. It also revealed a totally unexpected relationship between space and time. The three dimensions of space and the dimension of time became inseparable parts of something new, spacetime, a malleable framework that underlies the cosmos.

A few months later, as a dab of frosting on the relativity cake, Einstein showed that mass and energy can be converted into each other, following what is now the world's most famous equation, $E = mc^2$. Einstein realized that every bit of matter roils with immense energy, but it did not occur to him until many years later that this energy could be released. Neither he nor anyone else realized that the Atomic Age had dawned.

The revelations of 1905 proved a difficult act to follow. But Einstein accomplished that in his general theory of relativity, although it took him nearly a dozen years. He described the insight at the core of general relativity as the happiest thought of his life. It came to him while he was still working at the Bern patent office. A person in free fall, he realized, would not feel his own weight. Conversely, a physicist in a sealed rocket couldn't tell if it was accelerating uniformly or sitting on Earth. Gravity and acceleration, Einstein realized, are one and the same.

Einstein worked on the problem for over a decade, while at the same time moving through four increasingly prestigious academic positions, in Bern, Zurich, Prague, and Berlin. He settled in Berlin just before the outbreak of World War I. To add to the tumult around him, he divorced Mileva and married a cousin, Elsa Löwenthal. He had fathered three children with Mileva—the first, Lieserl, before they were married. To date, no one knows what became of her; she may have been adopted or she may have died as a young child. One of his sons became a successful scientist, teaching at the University of California, Berkeley. The other developed schizophrenia and lived out his life in a Swiss sanitarium.

Neither war nor divorce deterred Einstein from his work. Once again a simple observation, in his hands, led to earthshaking consequences. He found that Euclidean geometry, which assumes that space is flat, did not match the new reality he was exploring. Luckily, a mathematician named Bernard Riemann had developed a form of geometry describing curved spaces—just what Einstein needed. After years of what Einstein described to a friend as "positively superhuman efforts," he published his theory of general relativity in 1916. In it Einstein redefined gravity as geometry. Every particle of matter and every bit of energy in the universe warped the space-time fabric around it. Objects were no longer drawn toward each other by a Newtonian force acting instantaneously across space. Rather they glided along the shortest possible routes through the curved space-time continuum. Matter warps space; warped space determines how objects move.

Scientists hold general relativity in awe, both as one of the foundation stones of today's understanding of the universe, and also as a unique expression of a transcendent mind at play. Martin Rees, a leading cosmologist, writes, "Einstein's intellectual feat was especially astonishing because, unlike the pioneers of quantum theory, he wasn't motivated by any experimental enigma." The predictions and implications from Einstein's general theory of relativity continue to unfold today: black holes gnawing away at the heart of most galaxies, gravity waves rippling through space, and wormholes that may join parts of the universe separated by vast stretches of space and time. Its reach extends from the Big Bang to the accelerating expansion of the universe scientists now believe they are seeing.

From general relativity came the prediction that made Einstein, the lover of solitude, into a world-famous celebrity—the bending of light. In his 1916 paper, Einstein showed that the Sun's mass would bend the light from distant stars by a tiny amount. A solar eclipse in 1919 gave astronomers a chance to test his theory. The results matched

Albert Einstein.

his predictions; matter did warp the space around it. To date, relativity has passed every test.

From 1927 on, Einstein cast himself in the role of devil's advocate, stubbornly resisting the rising tide of quantum mechanics, a field that he himself had helped create. He clung to the conviction that the theory's foundation—the ultimate unpredictability of quantum events—meant that it was at best an incomplete description of nature. Hence his famous claim that God does not throw dice. Einstein and his collaborators posed a series of challenges to quantum theory, in the form of "thought experiments" which sought to reveal inconsistencies in the theory. His chief adversary was Niels Bohr, the leading quantum theoretician. Bohr inevitably found the flaws in Einstein's challenges. Einstein and Bohr disagreed deeply, but at the same time found their dialogue extremely stimulating. In a letter to Bohr, Einstein said, "Not often in my life has a human being caused me such joy by his mere presence as you have done." Perhaps he saw in Bohr a reflection of himself, writing of him, "He is like an extremely sensitive child who moves around in this world in a sort of trance."

Einstein spent the last decades of his life in Princeton, at the Institute for Advanced Study. He struggled for years to develop a unified field theory—a framework that would unite gravitation and electromagnetism. He gradually became aware that the task was beyond his powers. "I won't ever solve it," he wrote in 1948; "it will be forgotten and must later be discovered again." He proved prophetic; a central focus of twenty-first-century physics is the search for an even grander unified theory that will show how the four basic forces and the entire panoply of particles and anti-particles unfold from a single source.

Einstein was more than a great scientist. Throughout his life, he spoke out for peace and human rights, often at risk to himself. His lonely idealism led him to argue against militarism as Germany girded itself for World War I, and to oppose both Nazism and Communism before World War II. He very publicly refused to return to Germany after the Nazis came to power in 1933. Although he signed the famous letter addressed to President Roosevelt that led to the atomic bomb, Einstein went on to advocate nuclear disarmament and international cooperation during the Cold War. He was particularly pained to see America, his adopted homeland, descend into the repression of the McCarthy years. The FBI reportedly compiled a 1,000-page dossier on him. "I have hardly ever felt as alienated from people as I do right now," he wrote in 1950. "Brutality and lies are everywhere."

Many people accused Einstein of being naive in his advocacy of idealistic causes. There's no doubt about his idealism, but he was far from naive. He faced and fought humanity's dark side—anti-Semitism, militant racism and nationalism, and personal and political tyranny—all his life. He knew from personal experience how quickly praise can turn to abuse, prejudice to genocide. He fought bigotry and hatred passionately, but never at the same level as the bigots. "Arrows of hate have been shot at me too, but they never hit me," he wrote, "because somehow they belonged to another world with which I have no connection whatsoever."

To those who knew him, Einstein was warm, caring, loved to laugh, and was eminently likable. At the same time, those closest to him were aware that he wasn't completely of this world; he could slip effortlessly into a realm of thought where he alone could go. "I am truly a 'lone traveler,'" he wrote, "and have never belonged to my country, my home, my friends, or even my immediate family, with my whole heart. In the face of all this, I have never lost a sense of distance and the need for solitude."

"I have finished my task here," Einstein said just before he died. In the end, the sailor's compass that so intrigued him as a child had guided him a profound new understanding of the nature of the cosmos.

25

Wegener Sets the Continents Adrift

If no one dared the impossible, there could be no greatness on Earth.

—Kurt Wegener

Some people have great physical courage, others great intellectual courage. Alfred Wegener (1880–1930), explorer and scientist, possessed both. He died a few days after his fiftieth birthday, struggling to return to base camp after a mission to re-supply his colleagues pinned down for the winter in the middle of the Greenland Icecap. His death reflected his life, which was one of daring, endurance, and the willingness to face great hardships in pursuit of scientific knowledge. He revealed his intellectual vision and daring most dramatically in his theory of continental drift. A youthful flash of insight, followed by years of painstaking research, eventually revolutionized the earth sciences. Just as Copernicus displaced the Earth from her static, central place, Wegener uprooted the continents and sent them sailing through a slowly yielding sea of basalt. In a stroke, he solved mysteries from a dozen fields of study, and stirred questions still being debated today.

As his critics loved to point out, Wegener was not a geologist. In his years at the Universities of Heidelberg, Innsbruck, and Berlin, he studied astronomy and meteorology. He was steady and determined later in life, but as a student in Heidelberg he managed to get himself arrested for running down the main street wearing nothing but a sheet. At heart as much an adventurer as a scientist, he set a world ballooning record with his brother Kurt in 1906, soaring over Germany and Denmark for fifty-two hours. Wegener chose meteorology because astronomy "offers no opportunity for physical activity." His meteorological studies certainly gave him the opportunity he sought—they took him on four explorations to Greenland and Iceland; from 1906 to 1908, 1912 to 1913, and, as expedition leader, in 1929 and 1930. He fought on the Western Front during World War I, where he survived gunshot wounds to the arm and neck. With typical self-discipline, he managed to continue his research even in the chaos of war.

In 1910 Wegener noticed, as many had before him, the remarkable fit between the Atlantic coastlines of Africa and South America. He found that the edges of the continental shelves—where the continents fall off into the deep ocean—matched even better. At the time, he speculated that the continents might have once been joined, but decided it was highly improbable. A year later, however, Wegener read a scientific paper describing identical fossilized plants and animals found on both continents. Biologists had known about these matching fossils for decades. Since everyone—biologists and geologists alike—assumed that the continents were fixed in place, the biologists assumed that vast tracts of land must have bridged the Atlantic, but had long since subsided beneath the waves.

The matching fossils made a deep impression on Wegener, but he immediately recognized that ocean-spanning land bridges were impossible. He knew that the continents were made of lighter material than the ocean floors. Continent-sized landmasses could not have disappeared without creating telltale gravitational anomalies. The biological puzzle could be solved, he realized, if the continents had been united when the fossils were deposited, and had later split apart. The fossils, he wrote, "provided immediately such weighty corroboration that a conviction of the fundamental soundness of the idea took root in my mind."

The key features of what he called "continental displacement," but which soon became known as "continental drift" were simple. The lighter continental blocks float like icebergs in a denser layer of material whose surface forms the sea floors. The continents move over long periods of time, as shown by biological, geological, and climatological evidence. The Atlantic Ocean was the clearest case—it started as a fissure that split the Americas from Africa and widened over millions of years. In the distant past, all of today's continents were massed together, forming a vast supercontinent that Wegener called Pangaea, meaning "All Earth." As a continent plows through the ocean floor, its

Alfred Wegener.

———◆———

leading edge crumples, forming great mountain chains such as the Andes. In its wake it leaves behind chains of islands; and when one continent-sized landmass collides with another, great folded mountains result. The titanic collision between India and Asia is still building the Himalayas.

Wegener first presented his theory publicly on January 6, 1912. While recuperating from his war wounds, he wrote the first edition of his work, boldly titled *The Origin of Continents and Oceans.* He published it in 1915, revised it several times over the years, and saw it translated into a half-dozen languages. In his eyes, continental drift provided a compellingly lucid explanation for a variety of mysteries— identical fossils and matching sequences of geological features on opposite sides of the oceans, the distribution of closely related present-day species, and widespread evidence that tropical forests once grew in today's arctic regions, while glaciers spread across today's tropics. His reconstruction of Pangaea, coupled with his equally radical idea that the Earth's Poles can migrate, placed the South Pole at the center of the mysterious tropical ice sheets, and showed the previously inexplicable subarctic coal beds neatly hugging the equator. He knew that continental drift was revolutionary, but thought that it was so compelling that it would gain rapid acceptance. "I do not believe the old ideas have more than ten years to live," he wrote to his father-in-law.

Despite his optimism, Wegener's radical new idea provoked enormous resistance. He was attacking the universally held belief in a stable Earth, plus he was a scientific outsider. Most geologists regarded him as an unwanted intruder. They clearly understood that if he was right, much of what they believed and taught was wrong. In 1926, an American geologist, R. T. Chamberlain quoted an unnamed colleague as saying, "If we are to believe Wegener's hypothesis, we must forget everything which has been learned in the past seventy years and start all over again." Since Wegener had brought together findings from many fields, specialists in those areas had little trouble finding bits and pieces of evidence to contradict him. And the theory was particularly easy to attack because neither Wegener nor anyone else at that time knew of a force powerful enough to move the continents against the enormous resistance of the underlying material. Prophetically, the British geologist Arthur Holmes proposed that radioactive decay deep within the Earth might generate enough heat to drive convection currents powerful enough to push the continents. But even Wegener was not convinced. In the final edition of his book, published the year before his death, he wrote, "The Newton of drift theory has not yet appeared."

Unlike many of the scientists who attacked him, Wegener insisted on keeping his eyes on the big picture—the Earth as a whole. "We are like a judge confronted by a defendant who declines to answer, and we must determine the truth from the circumstantial evidence," he wrote. "How would we assess an judge who based his decision on part of the available data only?"

It took fifty years for the revolution Wegener started to come about, in the modern theory of plate tectonics. By that time, scientists could probe the Earth with entirely new tools. In the 1950s, sensitive magnetometers let researchers map faint traces of the Earth's ancient magnetic field frozen into rocks as they formed, and radioactive dating let them trace geological changes through time. The result was a flood of undeniable evidence that proved Wegener's central insight— the continents do move. In the next decades, observers crisscrossed the oceans, mapping them in unprecedented detail, and measuring heat flow, gravity, and the transmission of seismic waves. By 1960, Henry Hess of Princeton was able to weave together a new synthesis: magma welling up along Earth's network of mid-ocean ridges creates new sea floor that spreads in both directions. The ocean basins and the continents ride on a dozen great plates. As Holmes had suggested, vast convection currents welling up from deep within Earth's mantle move the plates. At last, geology's Newton had found a force powerful enough to drive the continents.

Today, guided by the concept of plate tectonics, earth scientists can trace the tango of the continents back nearly a billion years. Using computers to speed up the dance a million billion times, the continents can be seen scurrying around the globe, clumping together and sailing apart like ice floes on a swirling river. As Wegener foresaw, the dance explains the distribution of fossils, minerals, mountain ranges, volcanoes, islands, and earthquakes in great detail. The Earth, it turns out, is far more dynamic than almost anyone, except for Wegener, dreamed. "There clearly remains but one possibility," he wrote, "there must be a hidden error in the assumptions alleged to be obvious. . . . The continents must have shifted."

In 1930 Wegener led an international scientific expedition to Greenland—his fourth arctic exploration. His goal was to make meteorological and geophysical measurements simultaneously at three sites along the 71st parallel. The Mid-Ice Station was located 250 miles inland, at an altitude of nearly 10,000 feet. The expedition's specially built motorized sleds failed, making it almost impossible to keep the station and its crew supplied. On September 21, Wegener and two of his colleagues set out to re-supply the station by dogsled. Temperatures reached sixty-five degrees below zero. The trip took forty days. Wegener reached the station, rested for one day, and then started back. Neither he nor his companion, twenty-two-year-old Rasmus Villumsen, survived. Earlier, Wegener had written to his brother, "What makes it easy for me to tolerate the many unpleasant aspects of daily life here is the great mission that must be completed." Before he could return to his home and family, he wrote, he had research to carry out and responsibilities to fulfill. "And then, thank goodness," he wrote, "the obligation to be a hero ends, too."

26

Hubble's Expanding Universe

[E]quipped with his five senses, man explores the universe
around him and calls the adventure science.

—Edwin Hubble

Hubble's law . . . , the straight-line relationship between velocity and distance, . . .
made as great a change in man's conception of the universe as the Copernican
revolution 400 years before. For, instead of an overall static picture of the
cosmos, it seemed that the universe must be regarded as expanding.

—G. J. Whitrow

As you read this, the Hubble Space Telescope is orbiting in the
blackness of space. From high above Earth's distorting atmosphere,
it continues to send us luminous images from the edge of space and
time. In its most penetrating searches, harvesting photons that have
been crossing the universe for billions of years, it has shown us ragtag
armies of proto-galaxies in the midst of the primordial clashes that led
to the more orderly cosmos we now inhabit.

In October 1923, Edwin Hubble (1889–1953), the astronomer
whose name the Space Telescope honors, was poring over photographs of the Andromeda Nebula taken through the 100-inch Hooker
Telescope atop Mount Wilson in southern California. Hubble was a
man to whom success seemed to come easily. Tall and athletic, he'd
been a star boxer and basketball player at the University of Chicago. It
was there that he got to know the astronomer George Hale. Hubble

was a natural choice for a Rhodes Scholarship to Oxford, where he chose to study law. He also won medals in track and boxed against the French champion, Georges Carpentier. During his years at Oxford, Hubble acquired a posh British accent and the confidence of a born leader. Bored by the law, he returned to astronomy, earning his Ph.D. in 1917. His observational skills won him a research position at the Mount Wilson observatory, where he would have access to the largest and most powerful telescope in the world. After volunteering to serve in the U.S. Army during World War I, Hubble went to Mt. Wilson and set out to study the nebulae—misty patches in the night sky whose nature remained a mystery.

Hubble had previously detected three stars that appeared brighter on the plate from the night of October 5, 1923 than in earlier photographs, and marked each with the letter N for nova. Now, by comparing a sequence of images, he saw that one of the three stars brightened and dimmed in a regular pattern. That marked it not as a nova, but as a Cepheid variable, a kind of star whose actual luminosity could be calculated from how long it took to brighten and dim. By measuring the ratio between the star's calculated luminosity and its apparent brightness, he could pin down its distance from Earth. So the discovery of that one Cepheid meant that he could measure the distance to the Andromeda Nebula for the first time. And that distance, he knew, would give him the answer to the great astronomical question of the day—whether Andromeda and other spiral nebulae were just patches of gas and dust within the Milky Way, or distant "island universes" in their own right.

We don't know just what Hubble felt at the time. The only clue is that, departing from his notorious reserve, he crossed out the letter N by a star in the upper-right corner of the plate, and wrote in capital letters, "VAR!" At that moment, the universe exploded in size and richness. The cosmos was no longer bounded by the Milky Way, estimated at the time to be about 30,000 light-years across. Rather, the Milky Way, our galaxy, was just one among millions of "island universes" in a truly vast cosmos.

The debate over the nature of the Milky Way dates back to the eighteenth century. The philosopher-scientist Immanuel Kant coined the phrase "island universe," and was one of the first to speculate that the nebulae—blurry patches in the night sky—were separate galaxies comparable to the Milky Way. But the French mathematician Pierre-Simon Laplace created a convincing theory that the nebulae were whirlpools of gas in the process of forming stars, and so were relatively nearby. By the end of the nineteenth century, the spiral arms that observers had found in many nebulae were seen as proof of Laplace's whirlpool model. It looked as though the Milky Way contained every object astronomers saw in the sky, that it comprised the entire cosmos.

The controversy had been spotlighted just four years before Hubble's discovery, in what has since been dubbed "The Great Debate." Harlow Shapley, one of Hubble's colleagues at Mount Wilson, squared off against Heber Curtis of the Lick Observatory at a meeting of the National Academy of Sciences on April 26, 1920. George Hale, founder and director of the Mount Wilson observatory, organized the debate, entitled "The Scale of the Universe." Hale was determined to use the 100-inch reflector to survey the cosmos with the aim of determining the Sun's place in the Milky Way and uncovering the true nature of the nebulae.

In his presentation, Shapley marshaled the evidence against the existence of "island universes." Based on his studies of Cepheid variables in globular star clusters, which he correctly associated with the Milky Way, he was able to determine that the Earth lay far from the galaxy's center. But factors he was not aware of, such as the dimming of light by dust within the galaxy led him to overestimate the size of the Milky Way. Correspondingly, he greatly underestimated the size of the nebulae, and their distances from the Earth. "I prefer to believe that they are not composed of stars at all," he concluded, "but are all truly nebulous objects."

Curtis countered with an impressive summary of the evidence that Andromeda and other spiral nebulae were siblings of the Milky Way, separated by vast distances. But even he greatly underestimated how far away they were, leading him to favor estimates of the size of the Milky Way that we now know were far too small. While Shapley took the absence of spiral nebulae in the plane of the Milky Way as evidence that they were somehow associated with it, Curtis argued just the opposite. He correctly guessed that our view of the distant universe was blocked by dust within the lens-shaped Milky Way. "The spirals are individual galaxies, or island universes," he concluded, "comparable with our own galaxy in dimensions and in number of component units."

Although Curtis was seen as having carried the debate, the question could only be settled through observations. The 100-inch telescope at Mount Wilson was the only instrument powerful enough to detect Cepheid variables in a distant galaxy. That's why Hubble had been photographing the Andromeda Nebula, its neighbor M33, and a third nebula, NGC 6822, night after night until his October 1923 discovery.

Henrietta Leavitt at the Harvard College Observatory had pinned down the relationship between the period and brightness of Cepheid variables in 1912. Using her formula, Hubble placed what he now knew to be the Andromeda Galaxy about a million light-years from Earth. That estimate, which turned out to be too small, still pushed it far beyond Shapley's grandiose estimate of the reach of the Milky Way. Hubble's paper on his discovery of Cepheids in spiral galaxies was read at a meeting of the American Astronomical Association in Washington, D.C., on New Year's Day, 1925. It marked the end of the centuries-old debate about the nature of nebulae, and flung open the door to our current understanding of our place in the cosmos.

Edwin Hubble let us see that the Milky Way is just one among uncountable numbers of galaxies. Through a decade of observation and measurement, he went on to find that those galaxies are all moving away from each other, at speeds that are proportional to their distances. Hubble's data forced another shift in human thought—the realization that the universe is not static but expanding. Other scientists soon realized that tracing that expansion back in time led to a moment when all the matter and energy in the universe were concentrated in one place. The Big Bang, as the astronomer Fred Hoyle dubbed the moment of creation, remains the best explanation of the universe we see today. Modified to include a flash of hyper-expansion, or inflation, an instant after the birth of the universe, the theory has passed every observational test so far. Among other things, it explains the distribution of galaxies and galaxy clusters, the amount of hydrogen and helium in the universe, and the uniform fog of microwaves permeating the universe—the fading echo of creation.

Today, cosmologists are exploring the possibility that our entire universe is just one of many that bubbled up spontaneously from the mysterious quantum realm. Whether there will be another Hubble, poring over some yet-to-be-captured image, who proves or disproves the existence of those "multiverses" remains to be seen.

27

Out of Africa

It was in 1929 that I first met you and your Taungs baby, Australopithecus, in your laboratory. Then you alone had the clarity of vision to diagnose it as an anthropoid form closer to mankind than any primate hitherto known. . . . You, like myself, have been the herald of perspectives to which the future only gives its assent step by step.

—Henri Breuil

Usually, what helped me most was the general agreement of a lot of other people that I was on the wrong track! . . . it generally proved valuable to explore the reverse of the accepted view.

—Raymond Dart

Few subjects are more fascinating, more mysterious, or more contentious than our own origins. In 1924, a young anatomist named Raymond Dart opened a box of broken stones and picked out the fossilized brain and limestone-crusted face of what he came to believe was an ancient human ancestor. The Taungs Child, as the fossil came to be known, propelled Dart into the center of a controversy about the birthplace of humankind and what it means to be human. Today's scientists have hundreds of specimens to study, vastly more powerful ways to analyze them, and the ability to reconstruct the prehistory of our genes. Still, the roots and branches of our family tree, and the place of the Taungs Child and her australopithecine kin within it, are being fought over as fiercely today as they were seventy-five years ago.

Dart (1893–1988), the fifth of nine children, was born on his parents' cattle farm in Toowong, Australia, in 1893. He did well in school, and with the help of scholarships and prizes went on to study at the University of Queensland in Brisbane and the medical school at the University of Sydney. After serving as a medic in France during the last year of World War I, he studied at the University of London under two of the leading anatomists of the day. Probably because his maverick qualities were already evident, the only job he was offered was professor of anatomy at the new University of Witwatersrand in Johannesburg. He and his American wife, Dora Tyree, sailed for South Africa in December 1922 with far more trepidation than excitement.

Dart never lacked energy or imagination. While building the fledgling anatomy department, he encouraged his students to bring in interesting fossils and bones to broaden their grasp of comparative anatomy. A student, Josephine Salmons, presented the fossilized skull of a baboon from a limestone mine at Taungs. Intrigued, Dart asked an old miner for more samples. In the second of two boxes of fossils sent by the miner, Dart found a cranial endocast, or fossilized brain, which was surprisingly large. "Here . . . was the replica of a brain three times as large as that of a baboon and considerably bigger than that of any adult chimpanzee," Dart later wrote. "It was a big bulging brain and, most important, the forebrain was so big and had grown so far backwards that it completely covered the hindbrain." After ransacking the box, he discovered a matching chunk of stone still hiding the creature's face. Dart knew that he had made a great discovery. "Here, I was certain, was one of the most significant finds ever made in the history of anthropology."

Anthropology was still a very young science. Fifty-five years earlier, in his book *The Descent of Man*, Darwin had prophesied that Africa would prove to be the cradle of mankind. In the ensuing decades, however, almost all human and pre-human fossils had been found in Europe or Asia. Despite Darwin's influence, the leading scholars of human prehistory firmly believed that Europe, with its Neanderthal and Piltdown fossils, or Asia, where Eugene Dubois had found the famous Java Man (now known as *Homo erectus*), was mankind's nursery.

It took Dart seventy-three days of patient work, much of it with his wife's knitting needles, to free the face from its stone mask. Finally, on December 23, 1924, the rock split away. "What emerged was a baby's face, an infant with a full set of milk (or deciduous) teeth and its first permanent molars just in the process of erupting," Dart wrote. "I doubt if there was any parent prouder of his offspring than I was of my 'Taungs baby' on that Christmas of 1924."

Raymond Dart holding the Taung skull.

If 1924 was his year of discovery, 1925 would prove to be the first of many years of controversy. Dart published his findings in the prestigious journal *Nature* in February. He detailed the creature's relatively large and advanced brain, its flattish face, lack of eyebrow ridges, high forehead, and humanoid teeth and jaw. And, he emphasized, the location of its foramen magnum—the opening where the spinal cord enters the base of the skull—showed that it was bipedal. Dart's Taungs Child walked upright, leaving its hands free to be used as "delicate tactual organs" and "instruments of its growing intelligence."

Dart believed that it was no accident that this ancient ancestor had evolved on the South African veldt. In the tropics, he wrote, "nature

was supplying with profligate and lavish hand an easy and sluggish solution" to the problem of survival. "For the production of man,"
he concluded, "a different apprenticeship was needed . . . where competition was keener between swiftness and stealth." It was southern
Africa, he believed, with its scarcity of water and "fierce and bitter
mammalian competition," that was the foundry in which nature forged
humanity.

Dart named his ancient creature conservatively enough, dubbing
it *Australopithecus africanus*, or Southern Ape of Africa. But he made
it clear that he believed it to be far more than just an ancient ape. "It
is obvious," he wrote, "that it represents a fossil group distinctly advanced beyond living anthropoids in those two dominantly human
characters of facial and dental recession on one hand, and improved
quality of the brain on the other . . . just those features, facial and
cerebral, which are to be anticipated in an extinct link between man
and his simian ancestors." It was obvious to Dart, although not to anyone else, that it was an ancient ancestor of mankind.

To Dart's dismay, both the public and the anthropological establishment turned on him and his baby. In England, the Taungs Child
was soon featured in songs and cartoons as the epitome of ugliness.
Dart received letters accusing him of being an instrument of the devil
and warning that he'd roast in hell. More seriously, the leading
anatomists in England and Europe united in criticizing Dart's conclusions. They believed that human evolution had started with the brain.
The Taungs Child reversed that assumption—it was bipedal, but with
a smallish brain. Some classified Dart's discovery as an early chimpanzee, others as an early gorilla, but none accepted it as a human ancestor. Dart later denied that the tidal wave of criticism, followed by
years of indifference, discouraged him. But in fact he retreated from
anthropology and focused his energy on the medical school. He did
not venture back into the field for nearly twenty years.

The man who lured him back into was Robert Broom. Broom was
a retired Scottish physician with a passion for fossils. Soon after Dart
discovered the Taungs Child, Broom burst into his laboratory and knelt
down "in adoration of our ancestor." Broom published an early defense
of Dart's findings, and set out to find more fossils to support them. Excavating at other limestone caves in the region, he soon found fossils
belonging to a larger-jawed creature which he called *Paranthropus robustus* (now *Australopithecus robustus*). Broom's indefatigable excava-
tions and advocacy eventually convinced most anthropologists that the

australopithecines were human ancestors. By 1948, Arthur Keith, one of the early critics of Dart's work, would write, "of all the fossil forms known to us, the australopithicinae are the nearest akin to man and the most likely to stand in the direct line of man's ascent." It's now thought that as many as seven species of australopithecines thrived across Africa between four and one million years ago, and that one of them founded the genus *Homo*, of which we are the only surviving species.

However, just as he was gaining a degree of scientific respectability, Dart created a new controversy. Excavating at a mine at Makapansgat, he found bones of another australopithecine which, from their dark color, he believed had been burned. Associated with them, he also found collections of bone fragments which he convinced himself had been used as tools and weapons. He named his new find *Australopithecus prometheus* after the mythological figure who gave mankind fire.

Maybe it was Dart's hardscrabble childhood, his battlefield experiences, or his view of Darwinian evolution as a fierce and bloody contest. For whatever reasons, in his mind's eye, these close cousins of his Taungs baby turned into fierce predators. Dart, more than anyone else, legitimized the idea that our australopithecine ancestors were avid hunters and natural-born killers. His ideas inspired Robert Ardrey to write his best-selling book *African Genesis*, which made the "killer ape" part of popular culture. The opening scene of the movie *2001: A Space Odyssey*, in which a man-ape smashes bones in a delirious celebration of new-found power, comes straight from Dart.

Dart's beliefs that *A. prometheus* controlled fire, made tools of the bones, teeth, and horns of animals, and used some of these as weapons to hunt large animals, have not stood the test of time. The bones turned out to have been darkened by a natural mineral, and it's now generally accepted that significant tool use began not with the australopithecines but with their likely descendants, the earliest members of our own genus *Homo*. Mary Leakey's *Homo habilis*, or Handy Man, typifies these larger-brained and more versatile ancestors. Dart's emphasis on hunting as the driving force behind humanity's evolution has also stirred more disagreement than support. Most anthropologists now believe that cooperation was at least as important as aggression in our past, and that the patient gathering of food almost certainly nourished our ancient ancestors far more than big-game hunting.

However, Dart's most profound conclusions, that Africa was the cradle of humanity, and that an upright posture preceded a big brain, remain sound. Most anthropologists now believe that australopithecines were our ancestors. Still, human origins remain contentious. The discovery of the so-called Millennium Ancestor from Kenya, could extend humankind's roots two million years before the first australopithecines, and, conceivably, push them to the periphery of our family tree. However this current controversy plays out, Dart will remain as the first to validate Darwin's belief in humanity's ancient African roots.

More than that, writes Ian Tattersall, the head of anthropology at the American Museum of Natural History, Dart's vivid imagination breathed life into the stones and bones that preserve our history. "Dart was the first student of the human fossil record who seems to have felt viscerally the drama of the biological history of mankind," he writes. "He was certainly among the first to perceive that the story of our species is one of individuals, populations, and species living, striving, and dying as part of an enormously complex web of organic life, and to see that this story is not complete if we don't let ourselves go well beyond the teeth and bones that are our primary evidence of it."

28

Fermi and the Fire of the Gods

The Italian navigator has just landed in the New World.

—*Arthur Compton, in a coded message*
confirming the first sustained nuclear reaction,
December 2, 1942

Just before dawn on July 16, 1945, the brightest light ever seen on Earth seared the mountains and sky surrounding the stark Jornada del Muerto in southern New Mexico. An all-engulfing sphere of fire quickly transformed itself into a luminescent purple bloom roiling up through the clouds. Culminating an unprecedented crash program, a huge team of scientists and engineers had detonated the first atomic bomb, ushering the human race, for better or worse, into the atomic age.

Nine miles from ground zero, Enrico Fermi (1901–1954) and his fellow physicist Emilio Segrè watched the seething fireball. Segrè later wrote, "For a moment I thought the explosion might set fire to the atmosphere and thus finish the Earth." But Fermi, characteristically, was absorbed in an experiment. As the blast wave thundered by forty seconds after the detonation, he dropped scraps of paper from his hand. The blast deposited a few of them eight feet away. Fermi, as usual, was ready—he'd compiled a table that would immediately tell him how powerful the explosion had been. So he was the first to know that a few pounds of plutonium had released more energy than 10,000 tons of TNT. A few weeks later, the United States would devastate

Hiroshima and Nagasaki with atomic bombs, forcing Japan to surrender unconditionally. The Bomb brought a quick end to World War II, but put the entire human race under the long shadow of nuclear war.

Fermi had not set out to build a bomb. As a child growing up in Rome in the early years of the twentieth century, he simply found himself drawn to science. The third child of intelligent and capable parents, he stood out because of his phenomenal memory for poetry. He did well in all school subjects, but towered over his fellow students in math. Although he loved the outdoors, and throughout his life hiked, skied, climbed mountains and swam, he spent hours poring over physics books that he tracked down among Rome's used book stalls. A friend of his father who was a university-trained engineer gave Fermi carefully chosen and increasingly difficult scientific books to study. Fermi quickly read them, mastered them, and committed them to memory. Fermi also had a strong practical side—as a child he built electric motors, and as a teenager he made the apparatus for his own experiments. When he took the examination to enter the Scuola Normale Superiore, an elite school in Pisa, the evaluator took him aside to tell him that he was unique, extraordinary, and destined to become an important scientist.

Fermi was so gifted in fact, and so matter-of-fact about his gifts, that his fellow students and even his teachers soon looked to him as an authority. At his teachers' request, he brought them up-to-date about relativity and quantum theory. By the age of twenty-two, immediately after earning his doctoral degree from the University of Pisa, he was already seen as the person who would renew Italian physics. He was asked to write a chapter for a book on relativity. Prophetically, in his essay "Mass in Relativity Theory," he identified the possible release of nuclear energy as relativity's most important implication.

Fermi continued his research, studied at the centers of physics in Germany and Holland, and began to teach. At the age of twenty-six, he became the first professor of theoretical physics at the University of Rome. After squiring the smart and attractive Laura Capon around Rome in his bright yellow Peugeot convertible, he married her in July 1928. Despite Fermi's intense devotion to his work, their marriage was a happy one, and soon produced two children, Nella and Giulio.

Fermi's first great accomplishment flowed from Wolfgang Pauli's 1925 discovery of the exclusion principle. Pauli showed that the rules of the quantum world dictated that certain kinds of particles could not, in effect, occupy the same place at the same time. Fermi, who had tried unsuccessfully to explain the behavior of an idealized gas composed of identical particles, saw that this was the key he needed. Within a few months he worked out the statistics of such a gas, as the English physicist Paul Dirac did a few months later. The resulting Fermi-Dirac statistics are basic to understanding electrons, protons, and neutrons—the building blocks of matter now known as fermions. They parallel the Bose-Einstein statistics that govern particles such as photons that can overlap in space and time, known as bosons. The work catapulted Fermi into the forefront of theoretical physics.

Fermi soon turned his attention from the atom to its nucleus. At the time, one of the unsolved problems of physics involved beta decay—the emission of an electron from a nucleus. It appeared to violate two of the most basic assumptions of physics, the conservation of energy, and the conservation of momentum. In 1930, Pauli predicted that something was carrying away the missing energy, a ghostly particle that Fermi soon dubbed the neutrino—"little neutral one" in Italian. Fermi realized that the problem required not just a new particle, but a new kind of force—the weak interaction—typified by a new fundamental constant, now known as the Fermi constant. The famous British science journal *Nature* rejected Fermi's paper on the subject as "too remote from physical reality." We now know that the weak interaction, along with gravity, electromagnetism, and the strong nuclear force, defines reality; these four forces allow matter and energy to exist and interact.

Fermi next turned to a subject that would fascinate him the rest of his life—the neutron. Scientists such as Irène Joliot-Curie and her husband, Frédéric, had induced artificial radioactivity by bombarding substances with alpha particles, or helium nuclei. Fermi believed that neutrons would make far better probes of nuclei because, unlike the positively charged alpha particles, they were electrically neutral. They could slip through the wall of positive charge surrounding the nucleus and dive into its heart. Fermi now had the chance to prove himself as an experimenter. He and the young researchers who had flocked to work with him in Rome proceeded systematically. They started with the lightest element, hydrogen, and worked their way up to the most massive, uranium, bombarding each with neutrons from a glass tube filled with radon gas and powdered beryllium. Despite operating on a shoestring budget, they produced and identified a slew of new radioactive isotopes and, they thought, a new element heavier than uranium. But that was only the start.

The team was baffled by a mystery. They got very different results, depending on where they worked. Segrè writes, "There were certain tables . . . which had miraculous properties, since silver irradiated on these tables became much more active than when it was irradiated on other, marble tables in the same room." It was Fermi's unconscious mind that solved the problem. It led him "spontaneously" to place a lump of paraffin—ordinary candle wax—between their neutron source and their target, rather than the carefully machined wedge of lead they had planned to use. The neutron beam suddenly sparked a hundred times more radioactivity in the target than they expected.

Typically, Fermi called for a lunch break. By the time his crew returned, he knew the answer: hydrogen atoms within the paraffin must be acting like blockers on a football team, slowing the neutrons down. And, against all expectations, the target nuclei absorbed far more of these lumbering, low-energy neutrons than speedier ones. Hydrogen or carbon atoms in the "miraculous" wooden table must have produced the same effect. So, they soon found, did water. He knew they had made a great discovery, but he did not yet realize that he had found the key that would unleash the enormous energy Einstein had shown was locked up within all matter. The discovery won Fermi the Nobel Prize in 1938, and determined his course for the rest of his life.

Leo Szilard, a brilliant Hungarian physicist, had caught a glimpse of the same key five years earlier, as he stood on a street corner in London. Far more politically aware and foresighted than most scientists, Szilard was among the first to flee Germany after Hitler took power. The idea that came to him was of a nuclear "chain reaction" that could release vast amounts of energy. He immediately realized the military potential of his insight, obtained a secret British patent on the idea, and assigned it to the British Admiralty. At the time, however, neither he nor anyone else knew of a substance in which a chain reaction could take place.

By the fall of 1938 Fermi could no longer stay in Italy. His wife was Jewish. The racial laws being enacted by Mussolini and his Fascist government threatened her and her family. Reluctantly, he and Laura made secret plans to flee to America immediately after the Nobel Prize ceremony in Stockholm on December 10, 1938. While Fermi was en route to America, Otto Hahn and Fritz Strassmann in Germany showed that uranium bombarded with neutrons produced a much lighter element, barium. Lisa Meitner and her nephew, Otto Frisch, both refugees from Nazi Germany, quickly showed that the captured

$$-\frac{\hbar}{i}\frac{\partial}{\partial t} = \frac{p^2}{2m} \bullet \frac{Ze^2}{r}$$

$$\alpha = \frac{\hbar^2}{ec}$$

Enrico Fermi.

neutron transformed the uranium nucleus from a stable to an unstable form. Overflowing with energy, the new nucleus split in two, releasing a blast of energy. Mankind had discovered nuclear fission.

Fermi and his family started their new life in America on January 2, 1939. "We have founded the American branch of the Fermi family," Fermi told his wife. The news that uranium atoms had been split reached him within days of his arrival in New York. He realized that his team in Rome must have split uranium years earlier but had lacked the chemical skills to recognize the lightweight fission products. Fermi, like Szilard, immediately realized that if a uranium atom released more than one neutron when it split in two it could lead to a chain reaction. Here was the key to the energy of the atom, if it could be made to work. But preliminary experiments by Fermi, Szilard, and other scientists around the world indicated that it would not be easy to create a chain reaction—a way would have to be found to capture just about every neutron produced.

It was Fermi who found the way. Working first at Columbia University, and later at the University of Chicago, Fermi built the first atomic reactor. Assembled over the course of two months in an unused racquetball court under the grandstands of the University of Chicago's athletic field, the "pile" consisted of layers of ultra-pure graphite seeded with chunks of uranium. Following the Japanese attack on Pearl Harbor, on December 7, 1941, the project became government-funded and top secret.

In contrast to the tiny experiments which led up to it, the pile was enormous—200 tons of graphite bricks and 6 tons of uranium towering nearly to the ceiling of the ball court. Fermi's measurements and calculations told him that the pile was ready to sustain a chain reaction—"go critical"—the night of December 1, 1942. The next morning was spent inching the last cadmium-coated control rod out of the pile and monitoring the rising radioactivity. Never one to be rushed, Fermi called for a lunch break. When his team and the observers returned, he directed the control rod to be pulled out one more foot. The radiation counters leaped, then stabilized, showing that a self-sustaining nuclear chain reaction was taking place. Fermi had tapped the power of the atom.

Spurred by Szilard, who had been desperate to prevent Nazi Germany from beating the Allies to the atomic bomb, Fermi was the first scientist to try to alert the U.S. government to the risk. It was Fermi's work at Columbia that Einstein, also pushed by Szilard, cited in his famous August 2, 1939 letter to President Roosevelt. Pilot projects at Columbia, Berkeley, and elsewhere were gathered together under government control, leading to the two-billion-dollar, five-year Manhattan Project, and to the atomic bomb.

Fermi continued to play a central scientific role throughout the war years. And when the war was over, he allowed himself to be pulled away from his laboratory to serve on key committees advising the government on the future of nuclear research. He had always seen secrecy as the enemy of science, and now became one of the strongest advocates for open, international cooperation concerning atomic energy. Despite his role in creating the atomic bomb, he strongly opposed Edward Teller and others pushing to develop the far more powerful hydrogen bomb. "It is necessarily an evil thing considered in any light," he wrote. Fermi died of cancer on November 29, 1954, at the age of 53.

The atomic bomb was used, Winston Churchill wrote, "to avert a vast, indefinite butchery" and it seemed to him and many others outside of Japan, "a miracle of deliverance." Despite a massive buildup of nuclear weapons on both sides of the Cold War, and the possession of similar weapons by an increasing number of countries, nearly six decades have now passed without a repetition. Fermi, Szilard, and many of the other scientists who gave mankind the power to split the atom fervently hoped that it would make war impossible. They had witnessed the alternative.

29

McClintock's Chromosomes

There has never been anyone like Barbara McClintock
in this world, nor ever will be.

—*Howard Green*

One of the qualities that so impressed those who knew Barbara McClintock (1902–1992) was her vision. Her ability to see, to observe, and to understand the most subtle and complex genetic and developmental processes was not just unsurpassed, but to many of her peers, incomprehensible. Describing some work she did at Stanford analyzing for the first time the structure of the minute chromosomes of the mold *Neurospora*, McClintock told her biographer, Evelyn Keller, "I found that the more I worked with them the bigger and bigger [they] got, and when I was really working with them I wasn't outside, I was down there. I was part of the system. I was right down there with them, and everything got big. I even was able to see the internal parts of the chromosomes—actually everything was there. It surprised me because I actually felt as if I were right down there and these were my friends."

It should not come as a surprise that many scientists could not or would not follow where her vision led. Joshua Lederberg, who, like McClintock, would go on to win the Nobel Prize, announced, "By god, that woman is either crazy or a genius." She is now acclaimed as a genius, a scientist whose unique insight and impeccable research allowed her to see, decades before anyone else, that the genetic systems

162

of maize, molds, and mankind are not linear, unchanging, read-only texts, but flexible, dynamic systems, able not only to orchestrate the development of an organism, but to reorganize themselves.

Some of what made McClintock unique appeared early in her life. She was the third daughter of parents who valued independence and individuality. From infancy she showed a remarkable ability to occupy herself. She became even more independent after being sent to live with an aunt and uncle during her preschool years. Encouraged by her father, she became a tomboy, choosing tools over dolls and joining neighborhood boys in rough-and-tumble play. But she also spent hours absorbed in her own thoughts. She experienced her drives and feelings intensely. Her mother had to stop Barbara's music lessons because she threw herself into them with such intensity. In her teens, she developed a passion for learning, and especially for solving problems—her own way. "It was a tremendous joy, the whole process of finding that answer," she told Keller, "just pure joy."

McClintock was determined to go to college, but her mother was opposed to the idea. However, when her father returned from serving as a military surgeon in World War I, he sided with Barbara. At Cornell, she threw herself into her studies with her usual intensity, but also dated, played banjo in a jazz band, and was rushed by a sorority (although she turned them down). She was even elected president of the freshman class. But as her interest in science grew, she pulled back from other involvements. By the time she graduated in 1923, she had decided that a close relationship was not for her. "These attachments wouldn't have lasted," she told Keller years later. "I was just not adjusted, never had been, to being closely associated with anybody, even members of my own family."

McClintock dates her fascination with the young field of genetics from the fall of 1921, when she took the only undergraduate course in genetics at Cornell. Her teacher, C. B. Hutchison, spotted her talent. When the class ended, he invited her to take a graduate course in genetics. "Obviously, this telephone call cast the die for my future," McClintock later wrote. She also took a course in cytology, the microscopic study of cells. She soon became remarkably adept at visualizing and understanding the inner structures of cells and their chromosomes—which had only recently been proven to carry life's "heritable factors." That combination—cytogenetics—became her life's work.

The first years of her career were stellar. Using a new staining technique, she identified the ten chromosomes of the corn plant, *Zea mays*, and described them in unprecedented detail. She began to match specific characteristics of corn plants with particular markers on the chromosomes. Between 1929 and 1931, she published nine important papers on corn genetics. This burst of productivity culminated in a seminal paper in the *Proceedings of the National Academy of Sciences* in 1931. McClintock and her student, Harriet Creighton, proved for the first time that the appearance of traits inherited from both of a plant's parents was caused by the physical "crossing over" of chromosomal material during the formation of reproductive cells. It was a cornerstone contribution, writes James A. Peters in his *Classic Papers in Genetics*, the final link in a chain proving that chromosomes carry the genes.

For a few years at Cornell, McClintock enjoyed a kind of intellectual camaraderie that would prove all too rare in her life. Graduate students Marcus Rhoades and George Beadle came to study with Rollins A. Emerson, the leading corn geneticist of the day. McClintock, Beadle, and Rhoades would all make outstanding contributions to genetics, with McClintock and Beadle eventually winning Nobel Prizes. McClintock thrived on the daily contact with friends and colleagues who understood and valued what she was doing. "The communal experience profoundly affected each one of us," she wrote later.

McClintock had won early recognition for her discoveries. Yet, as a woman in science, no clear career path was open to her. Most American colleges simply didn't hire women as professors, except in fields such as home economics. Rhoades, Beadle, and other less accomplished men quickly moved on in their careers while McClintock struggled. For two years she continued her research, supported by a National Research Council fellowship. In 1933, she received a Guggenheim Fellowship to study in Germany, but she was traumatized by what she saw there as the Nazis came to power. She describes being "crushed, disturbed, and utterly panicked." She fled back to Cornell before the year was out.

Worried that McClintock might not be able to continue her research, Emerson got McClintock a small grant from the Rockefeller Foundation, which carried her through 1935. Finally, Lewis Stadler, a geneticist who admired her work, offered her an assistant professorship in his new program at the University of Missouri in Columbia, Missouri, where McClintock remained for five years. Although she continued to produce trailblazing research, it became obvious to her that she was tolerated rather than valued. The eccentricities of a brilliant male scientist might have been overlooked, but she was seen as a troublemaker. The University eventually showed some interest in her, but only after she was nominated to the prestigious National Academy of Science. But McClintock had already decided to leave. She eventually found a research position at the Cold Spring Harbor laboratory of the Carnegie Institution of Washington. That secluded setting on the North Shore of Long Island became her laboratory and home for fifty years, from the spring of 1942 until her death.

McClintock started her most important work in 1944. She noticed that some kernels of hybrid corn showed patches of color of different sizes. To her trained eye, these had to be caused by what she called "mutable genes," genes whose function seemed to be lost and regained as a plant developed. McClintock set aside all other work to track down what was happening in the chromosomes to produce those changing patterns. It took her two years to make her first breakthrough. She found a section of the chromosome that could break free from its normal location, and she realized that this dissociation depended on a neighboring gene, which was controlled by yet a third gene—the activator—located far away on the same chromosome. McClintock had discovered a mechanism in which one part of the genome reorganized another in a controlled way. But this was at a time when the idea that the genetic system might change or regulate itself was unheard of.

McClintock spent the next several years intensively studying dissociation and the genes that controlled it. By 1948, she had found that both the controlling element and the dissociated element reappeared in new locations on the chromosome. In other words, a piece of a chromosome could be given its marching orders, and would obediently move to an entirely different neighborhood. She called this entirely novel process transposition.

By 1949, McClintock was starting to see the workings of the genetic system in ways that the scientific community was simply not ready for. It was obvious to her that genes could cause other genes to move from place to place—and that when a gene inserted itself back into the chromosome, it might turn a nearby gene on or off. She knew that this process could have profound effects on an organism—a transposition early in development might produce major changes in the growing plants that she studied. The same transposition later in development might produce much smaller or more localized changes. These mobile genetic elements, she thought, could shed light on the mystery of how a single cell differentiates into all the different kinds of cells and tissues that form an organism, and on the process of evolution itself. None of this was obvious to others.

When McClintock presented her findings at one of the renowned summer symposiums at Cold Spring Harbor, the reaction of other geneticists was devastating—silence, snickering, mumbled complaints. For the next five years she published papers and gave talks about her discoveries, but encountered only indifference and hostility. A leading geneticist told her, "Now, I don't want to hear a thing about what you're doing. It may be interesting, but I understand it's kind of mad." McClintock had gotten used to personal isolation, but now she was professionally isolated too. She was deeply hurt, but never doubted what she'd found. She responded by becoming even more absorbed in her work. But for many years she stopped presenting her findings in person, and published her results only in the laboratory's annual reports.

Looking back, it's not difficult to find reasons for McClintock's descent into scientific solitary confinement. Many people had come to see her as a difficult person, and her work as meticulous but hard to understand. McClintock was never afraid to ask big questions, and her unique skills led her to radical new answers. Her assertion that genes could control each other clashed with a central belief that genetic variation must be random. Perhaps most importantly, her work crashed head-on into what was becoming the dominant force in the study of life, molecular biology.

In 1944, Oswald Avery and his colleagues showed that DNA appeared to carry genetic information. By 1952, Al Hershey and Martha Chase at Cold Spring Harbor found that DNA did that without help from surrounding proteins. In 1953, Watson and Crick deciphered DNA's exquisite double helix. It soon seemed that all the important questions in biology could be answered by reading the new-found code of life. And Crick enshrined as the central dogma of molecular biology the axiom that information could flow from genes into cells, but never back to the genes. In the brave new world of molecular biology, McClintock's vision seemed at best heretical, at worst, irrelevant.

For more than a decade, McClintock's colleagues responded to her work with icy silence. But then, step-by-step, a thaw set in. As molecular biology matured, it began to address the questions McClintock had been asking all along. It found patches of DNA that behaved exactly the way McClintock had described—soon to be dubbed "jumping genes." Grudgingly, molecular biologists began to realize that the genome, the totality of an organism's genetic machinery, does its many jobs through an elaborate system of variability and control. It's now known that in many organisms, including man, up to half of the DNA is in the form of transposable elements. In a recent article in the journal *Science*, Daniel Voytas concludes, "It was from McClintock's work with maize that we first learned of transposable elements and their ability to reorganize genomes. The wealth of maize retroelements further speaks to the profound fluidity of genomes and their abundant capacity for change." And, as McClintock predicted, those transposable elements play a vital role in both development and evolution.

After McClintock's visionary work was rediscovered, she was showered with awards, including the very first MacArthur "genius" Award and the 1983 Nobel Prize for Medicine or Physiology. Her reaction to acclaim was in keeping with her character. Her colleague Howard Green writes, "When transposons were demonstrated in bacteria, yeast, and other organisms, Barbara rose to a stratospheric level in the general esteem of the scientific world and honors were showered upon her. But she could hardly bear them. She felt obliged to submit to them: it was not joy or even satisfaction that she experienced; it was martyrdom! To have her work understood and acknowledged was one thing, but to make public appearances and submit to ceremonies was quite another."

Not despite her uniqueness, but because of it, McClintock has joined the rarefied company of great explorers, innovators, and discoverers—epoch-making scientists such as Pasteur, Mendel, Curie, Darwin, and Einstein. But she would be the first to announce, in no uncertain terms, that her goal was neither belonging nor fame, but simply to see and understand. "It's going to be marvelous," she told her biographer Evelyn Keller. "We're going to have a completely new understanding of the relationship of DNA, the cell, and the organism as a whole."

30

A Bit of Genius

[The] semantic aspects of communication are irrelevant to the
engineering problem. The significant aspect is that the actual
message is one *selected from a set* of possible messages.

—*Claude Shannon, 1948*

He's one of the great men of the century. Without him, none of the things we
know today would exist. The whole digital revolution started with him.

—*Neil Sloane, 2001*

I visualize a time when we will be to robots what dogs are to humans. . . .
I am rooting for the machines.

—*Claude Shannon, 1987*

Like many people, I've spent most of the day at my computer. I've
surfed the Net, searched the catalogs of three libraries, printed out
several articles, and stored others in my computer's memory. I've also
sent and received reams of e-mail and made dozens of cellphone
calls. Perhaps tonight my wife and I will rent a movie. Throughout the
world, hundreds of millions of people have been working or playing
in similar ways. Remarkably, not one of these activities would be possible were it not for the ideas of a brilliant, whimsical, curiosity-driven
man working on his own more than fifty years ago.

Claude Shannon's childhood could have been painted by Norman
Rockwell. Shannon (1916–2001) grew up in Gaylord, Michigan, a
town of three thousand, where his mother was the high school princi-
pal and his father served as probate judge. "If you walked a couple of

blocks, you'd be in the countryside," Shannon said. A streak of intelligence and inventiveness ran through the family—his grandfather, a distant cousin to Thomas Edison, was a farmer and inventor with several patents to his name. While still a child, Shannon turned a neighbor's barbed-wire fence into a private telegraph line to a friend's house half a mile away. As a lanky teenager, he earned money delivering telegrams and repairing radios. His older sister, who became a professor of mathematics, stimulated his interest in numbers by sending him puzzles to solve. He followed her to the University of Michigan, where he earned undergraduate degrees in electrical engineering and mathematics—dual interests that would guide him throughout his career. He went on to earn his Ph.D. in mathematics from the Massachusetts Institute of Technology at the age of twenty-four.

Shannon accomplished his greatest work before the age of thirty-two. In his master's thesis, published by the American Institute of Electrical and Electronic Engineers in 1938, he made the perfect match between electronic switching circuits and the binary logic created by the nineteenth-century British mathematician George Boole. Shannon showed that simple electronic circuits could carry out all the operations of Boolean symbolic logic. At the practical level, his insight transformed digital circuit design from an art to a science. More broadly, his proof that Boole's "laws of thought" could be embodied in electronic circuits opened the door to the world of digital computers. His paper, "A Symbolic Analysis of Relay and Switching Circuits," has appropriately been described as one of the most important master's theses ever written.

Ten years later, flowing from his work at the Bell Telephone Laboratories, Shannon and co-author Warren Weaver presented his next breakthrough—Shannon's brilliant physical-mathematical definition of information. The work, "A Mathematical Theory of Communication," is considered to be the Magna Carta of the Information Age.

Shannon developed his theory of information to tackle a practical challenge faced by radio and telephone engineers—how to pump as much "information" as possible through an intrinsically noisy wire or microwave channel. But until Shannon, nobody knew what information was. Shannon attacked and solved the question at its most basic level. Demonstrating his great ability to drill to the core of an issue, he brushed aside all the specific types of information—images, sounds, words, or whatever. Similarly, he freed himself from analyzing any particular way of transmitting information—sound waves in air,

Claude Shannon.

microwaves, telephone wires, etc. What happened, he asked himself, when information traveled from a sender to a receiver? His answer was profound. Information reduced uncertainty. The simplest form of information, the unit, atom, or quantum of information, was an answer to a yes/no question. Did the market go up or down? Shall we meet today or not? Hold a coin in your hand and you have uncertainty. Flip it and you gain exactly one unit—a bit—of information.

Linking information to uncertainty made information subject to the laws of physics and the mathematics of probability. Shannon equated information with the fundamental thermodynamic concept of entropy. As developed by Ludwig Boltzmann in the nineteenth century, entropy measures the amount of disorder, or uncertainty, within a physical system, such as the gas in a balloon. As defined by Shannon, information measures the amount of uncertainty within a set of possible messages. The physical nature of information has proved to be a crucial insight. For example, it explains why "Maxwell's Demon" can't violate the Second Law of Thermodynamics by sorting out fast- and slow-moving molecules. To do that would require information, and gathering that information would require at least as much energy as the demon could gain.

Once he had a firm handle on the physics and mathematics of information, Shannon derived a string of important theorems. He found that there is a clear limit to how much information can be pushed through any communications channel in a given time, a quantity now known as Shannon's channel capacity. He derived an absolute relationship between the range of frequencies available for carrying a message and the amount of information that can be pumped through—hence today's high-tech preoccupation with "bandwidth" and the billions of dollars companies are willing to pay for their share of it.

One of Shannon's most remarkable findings was that digital information can be transmitted perfectly—or at least with any desired level of accuracy. Since every medium inevitably adds "noise" to a signal, any analog process—whether it's recording music, printing a photograph from a negative, or broadcasting a radio program—degrades the information. A copy of a copy of a copy may be unreadable. But once the sound, image, or measurement is translated into bits, it's safe. Shannon proved that as long as the sender does not try to exceed the capacity of the channel, properly encoded information can be sent virtually error free, despite the channel's inherent noise. Using the concepts of redundancy and error correction that Shannon pioneered, information is now routinely stored, sent, read, and duplicated nearly perfectly. Your CD does not add noise to your music, your DVD does not add snow to your TV screen, and distance doesn't degrade the images that you download from a website in Japan, or that astronomers receive from a spacecraft orbiting Jupiter.

Shannon's information theory, wrote coauthor Warren Weaver, "is deep enough so that . . . it is dealing with the real inner core of the communication problem—with those basic relationships which hold in general, no matter what special form the actual case may take." In short, Shannon discovered that information is physical; it is subject to physical laws just as much as water flowing through a pipe or air through a turbine. Today it is obvious that any pattern—a movie, a concert, a map, a CAT scan of the human body—can be reduced to bits, then stored, manipulated, and transmitted at will. There were, of course, movies, recordings, telephones, and faxes before Shannon. But their underlying unity—that they are all information, and only information—was invisible until he showed us what to look for. Shannon and subsequent generations of scientists and engineers have mapped the laws governing information and exploited them to develop the information infrastructure that powers today's world.

Shannon's insights were significantly ahead of their time. His colleagues Shockley, Brittain, and Bardeen were still working on the transistor. Integrated circuits, microchips—in fact all of today's silicon-based, information-dense world—lay in the future. Bits have been organized into bytes, bytes by the billions into programs, operating systems, and entire digital environments. Within the last decade, scientists have discovered the qubit—a quantum bit capable of representing not just yes or no, but yes and no at the same time. Ultra-powerful quantum computers may soon send today's most powerful computers to the junkyard. Shannon showed the way. We're still exploring the world he discovered.

It's ironic that Shannon's ideas have proved to be of such world-shaping importance. Practicality was never his goal. He was a whimsical genius who was just as happy riding his unicycle down the corridors of the Bell Labs, building juggling machines, or inventing a rocket-powered Frisbee, as plumbing the depths of information theory. His maze-learning mouse and chess-playing machines were early and important contributions to the field of artificial intelligence, but one can only join him in laughing at the THROBAC computer he built to calculate in Roman numerals. Aided and abetted by his wife Betty, who bought him a top-of-the-line Erector set and his first unicycle, Shannon was driven by curiosity and fun, not by practicality or gain. "I've always pursued my interests without much regard to financial value or value to the world," he says. "I've spent lots of time on totally useless things."

31

The Dynamic Duo
of DNA

It has not escaped our notice that the specific pairing we have postulated immediately suggests a possible copying mechanism for the genetic material.

—*James Watson and Francis Crick*, Nature, *April 25, 1953*

I never met two men who knew so little—and aspired to so much.

—*Erwin Chargaff*

$\mathbf{S}$ aturday, March 7, 1953, is a date that may be remembered longer
than Pearl Harbor or Hiroshima. That's the day when James Watson
(1928–) and Francis Crick (1916–), working feverishly in their office
at Cambridge, first pieced together the structure of DNA, the molecule at the core of life. Wars and even the nations that fight them
come and go, but the discovery of the structure and function of the remarkable and beautiful molecule that stores the accumulated adap-
tive experience of life on Earth has changed humanity's fate forever.

They met during the first week of October 1951. Watson was a
brash, crew-cut young American who had gone to the University of
Chicago at fifteen, earned his Ph.D. at twenty-two, and, boosted by his
famous advisor, Salvador Luria, wangled a postdoctoral berth at Cambridge. Crick, at age thirty-five, had ambled into biology from physics
after a long wartime stint designing mines for the British Admiralty.
During his years at the Cavendish Laboratory at Cambridge, he'd
made no progress toward his Ph.D., but had earned a reputation as a

Rosalind Franklin.

loud, overly talkative, often annoying gadfly who gleefully explained other people's results before they could. The powers that be gave Crick and Watson an office to share, "so you can talk to each other and not disturb the rest of us."

And talk they did. Over the next months, they compared notes, argued, shared boisterous lunches at the Eagle, a nearby, nondescript old pub, and in the process forged one of the most fruitful partnerships in the history of science. Watson energized Crick, pushing him to focus his formidable intelligence and understanding of crystallography on the structure of DNA. Crick critiqued Watson, filled in crucial gaps in his knowledge, and helped him to winnow out the best of his ideas from the rest. Their motivations differed greatly. The more gentlemanly Crick thought they were both "mad keen to solve the problem." Watson, it's clear, was a "top gun," driven to beat the great Linus Pauling at his own game, and hot to win the Nobel Prize. But for sixteen erratic, brilliant months, Watson and Crick complemented each other like the engine and chassis of a racing car.

One thing they shared, they quickly discovered, was the conviction that DNA held the secret of life. They wanted to transform genes from abstract entities, "beads on a string," to real molecules. Many biologists still thought that proteins, not the seemingly simpler nucleic acids, choreographed the genetic dance. Watson and Crick, correctly interpreting the latest research findings, bet on DNA. They hoped that its structure would explain the vital functions of genes—their ability to replicate, to control cellular chemistry, and to carry specific traits from generation to generation.

Scoping out the structure of DNA was at the shadowy edge of scientific capability at the time. Clues lay scattered in the published literature and unpublished expertise of a score of scientists. In 1947, William Astbury had published one blurry photograph of X-rays diffracted by passing through DNA. A Norwegian student in London, Sven Furburg, took a few more diffraction pictures and correctly deduced that DNA's four bases, carbon-based ring-shaped structures, lay at right angles to a sugar-phosphate chain. He reported his results in his doctoral dissertation and went home. Buried in lengthy papers by Erwin Chargaff was the observation that, although DNA's four bases appeared in different proportions in different species, the amount of adenine always equaled the amount of thymine, and guanine always balanced cytosine. In April 1951, Pauling had shown that one of the basic structures of proteins was a spring-shaped molecule—the alpha helix. There were reasons to think that DNA took on a similar form. Reasons, but no details—and no proof.

The laboratory that had the best chance to uncover the crucial details was not the Cavendish at Cambridge, but the lab headed by John Randall at King's College in London. There, Rosalind Franklin and Maurice Wilkins had the funding and expertise to study DNA. But just as personal chemistry bonded Watson and Crick into a devastatingly effective scientific commando unit, chemistry of a different kind kept Wilkins and Franklin locked in their separate offices. They hardly spoke to each other, and managed to hinder rather than help each other's work. It was partly a gender battle. Wilkins, at the lab and working on DNA before Franklin arrived, expected her to help him. Franklin, however, did not report to him, and in fact had little respect for him. She was already an accomplished researcher, and was determined to do first-rate science in what was clearly a male domain. Their relationship quickly spiraled into mutual hostility.

Watson and Crick made their first stab at DNA toward the end of 1951. It was a disaster. After half-listening to Franklin presenting her preliminary findings at a seminar, Watson returned to Cambridge eager to crack DNA's structure. Relying on the few clues they had, he and Crick wired together a model. It sported three strands of DNA spiraling upward, with their sugar-phosphate "backbones" in the center and bases sticking out like the branches of a Christmas tree. It fit the little data they had. Exuberant, they invited Wilkins and Franklin to Cambridge. Franklin demolished their scheme in short order. Among other flaws, their molecule held less than one-tenth the amount of water she had found in DNA. Watson, who hadn't bothered to take notes at her talk, had missed that crucial point. He had also ignored her conclusion that the phosphate chains ran along the outside of the molecule.

The fallout from their botched coup was nearly fatal to their plans. Sir Lawrence Bragg, the head of the Cavendish Laboratory, reminded them that DNA was a King's College project by priority and funding. He ordered Crick and Watson to stay away from DNA. As Watson later wrote, "Lying low made sense," at least as much from the fact that they didn't know how to proceed as out of respect for Bragg's authority. They even gave Wilkins the forms used to build their model atoms and molecules, in the hope that he would use them to model DNA. Wilkins accepted the dies, but never used them.

Watson and Crick resumed their assault on the DNA molecule a year later. The catalyst was Linus Pauling's entry into the arena. Alerted by Pauling's son Peter, who had started as a graduate student in their lab, Watson and Crick felt they had no time to waste. Luckily, in the previous months they had picked up some vital clues. Crick had asked a young mathematician, John Griffith, to calculate how DNA's bases might bond chemically. Griffith found that adenine would bond with thymine and guanine with cytosine. Crick noticed, and filed away, that this might explain the ratios Chargaff had found, and realized that this molecular pairing might underlie genetic replication. Watson and Crick had come across work by Alexander Todd showing just how the sugar-phosphate chain was linked together chemically. They had also gained access to a memo, circulated by Franklin's laboratory for administrative reasons, which gave them a bit more of her data. Watson had done some X-ray work on his own, although not on DNA, which sharpened his understanding of diffraction photographs. When Wilkins showed him one of Franklin's best diffraction photographs, Watson was now able to recognize a striking helical signature, and guess at a crucial ratio. And, to their great relief, Pauling's DNA model was wrong. Just as they had done a year earlier, Pauling came up with a three-strand molecule with the sugar-phosphate backbones inside.

Early in February 1953, Watson and Crick met with Bragg and pleaded with him to end their exile. They argued that they would really be competing against Pauling in America, not Wilkins and Franklin in London. Bragg, who resented Pauling for having beaten him, years earlier, to the basic rules for interpreting X-ray diffraction results, unleashed them.

By now their teamwork was seamless. Watson built a two-stranded model, but once again with the backbones inside. Crick overcame Watson's resistance and got him to try them outside. They practically fell into place, but didn't leave enough room for the bases. Crick realized that a particular symmetry in Franklin's data implied that the two chains ran in opposite directions. That doubled the height of each spiral, providing crucial room for the bases. Impatient with the machine shop, Watson cut models of the four bases out of cardboard. Playing with them, he saw that adenine and thymine could bond naturally, and formed a shape that was identical to a guanine-cytosine pair. When he connected the paired bases to their phosphate backbones, they clicked into place like the rungs of a spiral staircase. This time everything matched—the physical locations of all the atoms, the chemical bonds, the end-to-end chains spiraling together, and the perfectly shaped pairs of bases. By the last Friday in February 1953, Watson and Crick had it, and they knew it. At lunch at the Eagle, they let anyone know who cared to listen that they had, putting it modestly, "discovered the secret of life." Gerard Pomerat of the Rockefeller Foundation was visiting the Cavendish Laboratory at that time. He described Watson and Crick as "somewhat mad hatters who bubble over about their new structure."

During much of 1952, Wilkins had put his own DNA work on the shelf. He told Watson and Crick that he planned to dive in again after Franklin left. Franklin struggled toward her goal without a colleague who could fill in gaps in her knowledge or wave her away from blind alleys. Early on she had found that DNA comes in two forms, which she labeled A and B. She quickly deduced that the B form had a helical structure. But she later wasted months exploring a possible figure-eight structure for the A form of the molecule. Still, despite working alone and in a hostile atmosphere, she came close to beating Watson and Crick to the structure of DNA. Her laboratory notes show that by the spring of 1953 she knew that DNA was a double helix, knew some of the molecule's key dimensions, knew that the phosphate scaffolding coiled around the outside of the molecule, and was starting to grapple with how the bases were packed inside.

The Watson-Crick partnership, which Crick described as "electrifying," soon unraveled. They published their key papers in *Nature* and in the *Proceedings of the Royal Society* before the end of 1953. Watson left for California, and from there fired a blast at Crick. He accused Crick of bad taste for considering an appearance on a BBC science program. "If you need the money that bad," he wrote, "then go ahead. Needless to say I shall not think any higher of you and shall have good reason to avoid any further association with you." They did appear together nine years later, along with Wilkins, to accept the Nobel Prize. Although Franklin's work had provided them with crucial data, she could not be considered for the Prize; she had died of cancer in April 1958. Any healing that time and honors may have produced was undone in 1968 when Watson published *The Double Helix*. In it, he depicted everyone but himself and Crick as bunglers, cast Crick as arrogant, and mercilessly caricatured Franklin. Still, in the end, the greatness of their discovery overrides all other issues. As Watson points out, "Francis Crick and I made the discovery of the century, that was pretty clear. We made it, and I guess time has justified people paying all this respect to me in spite of my bad manners."

Like any discovery, the structure of DNA marked the end of a long, often disjointed chain of progress. But far more than most, it also marked a beginning, not just of a new field of research, or even a new science, but of a new era in human history. As nearly fifty years of explosive study has shown, understanding DNA changed everything. Within a decade, it explained exactly how genes store information in a four-letter chemical alphabet, how they replicate, and how they drive not only the development and functioning of the cell, but heredity and evolution as well. An army of researchers has gone on to provide tools for deciphering the molecular mechanisms of the cell, the genetic history of the human race, and, in fact, of the branching and sometimes entangled tree of life itself. They have now copied out the entire genome of our own species and many others, and are beginning to interpret them, gene by gene, protein by protein, interaction by interaction. And what we can read we can also write. We have deliberately modified the heritable genetic code of many plants and animals, and, at least in one case, a human. We are learning how to feed more people, diagnose and cure devastating diseases, and turn organisms into factories for our own benefit. We may soon be able to extend the human lifespan, and even create entirely new kinds of organisms. Nobody can predict what the future of the human species and our fellow creatures will be. But, with the master-molecule of life now in our hands, it can never be the same.

32

Echoes of Creation

We had a list of projects that we were going to do, nothing that was going to get a tremendous amount of attention from the world.

—*Robert Wilson*

Measurements of the effective zenith noise temperature of the twenty-foot horn-reflector antenna at the Crawford Hill Laboratory, Holmdel, New Jersey, at 4080 Mc/s have yielded a value about 3.5 K higher than expected.

—*Arno Penzias and Robert Wilson*

Well, boys, we've been scooped.

—*R. H. Dicke*

N oise, not silence, has proved to be golden for radioastronomers. Karl Jansky, a radio engineer at the Bell Telephone Laboratory, founded the field of radioastronomy in 1932 when he found that an annoying hiss he'd been trying to track down for several years peaked every 23 hours, 56 minutes—precisely one day, not by clock time, but by star time. The mystery radio source turned out to be the center of the Milky Way. Jansky's discovery planted the seed of radioastronomy. It did not begin to grow until after World War II, although radio and radar engineers often had to work around "cosmic static" as they improved their antennas and receivers.

In the early 1960s, the Bell Telephone scientists Arno Penzias (1933–) and Robert Wilson (1936–) were trying to track down an even more mysterious hiss in the communications antenna they wanted to use as a radiotelescope. They found the unexplained buzz whenever and wherever they pointed the antenna. It didn't vary from night to day, from season to season, or from place to place in the sky. Much to their surprise, this "excess noise" turned out to be of fundamental importance—nothing less than a whisper from the dawn of time.

Penzias may be one of the greatest overachievers of all time. Born to Jewish parents in Munich in 1933, he grew up in a family struggling to survive Hitler's program of persecution and extermination. He and his brother, followed by their parents, escaped to England and went on to New York just before World War II. His father found work as a building superintendent, his mother in a garment factory. As a poor kid with a foreign accent, Penzias did not fit in at school. He finally made some friends at Brooklyn Technical High School, where he studied chemistry rather than his real interest, electronics. He switched to physics at the City College of New York, where he considered himself a barely adequate student. MIT turned him down, so he did his graduate work at Columbia, earning his Ph.D. in 1962. "I just got through Columbia by the skin of my teeth," Penzias says. The gap between his accomplishments and his assessment of his own worth persisted right up until he won the Nobel Prize. When he read an article predicting that he and Wilson would win the Nobel, he thought it so absurd that he laughed out loud.

Penzias took a job at the Bell Laboratories at Holmdel, New Jersey, in 1961. A second radioastronomer, Robert Wilson, arrived in 1963. They agreed to share the one radioastronomy position that was available. They made a good team, with the careful, quietly confident Wilson anchoring the more voluble and driven Penzias. Wilson came from an all-American background. His grandparents were Texas farmers, his parents the first in their families to go to college. In high school, Wilson played trombone in the marching band. Like his father, he enjoyed electronics and soon was fixing radios and television sets "for fun and spending money." At Rice University, in Houston, his talent for physics and math became evident, allowing him to chose between graduate programs at MIT and Caltech. It was at Caltech that he became interested in radioastronomy. His research in that area led to the job at Bell Labs the year after his Ph.D.

At the time, Bell Labs owned a unique instrument—a huge, horn-shaped antenna that had been built to detect microwave signals bounced from ECHO, a giant reflective balloon launched in August 1959 to demonstrate the feasibility of communication satellites. By the time Wilson arrived, Bell was leaving the communication satellite business, so Wilson and Penzias were able to get permission to convert the antenna into a radiotelescope. They planned to study radio emissions from the Milky Way and from a supernova remnant called Cassiopeia A.

Before they could start they had to calibrate their receiver. They tried to do this by comparing the quietest signal from the antenna to a "cold load," a signal from a source chilled by liquid helium to within a few degrees of absolute zero. To their dismay, they found that the antenna produced an inexplicably noisy signal. In radio engineering terms, it was as if the antenna were inside a giant box warmed 3.5° to 4° above absolute zero. If they could not account for the noise, they could not trust the observations they hoped to make.

The efforts to track down the noise source made by Penzias and Wilson over the course of the next year have become legendary. They found a pair of pigeons nesting in the antenna. But removing them (twice) along with the "white dielectric material" the pigeons had deposited made no difference. They covered aluminum rivets with special tape. They ruled out stray signals from an atmospheric nuclear test, from the atmosphere itself, and from New York City. They rebuilt part of the horn because they thought its shape wasn't quite right. They tinkered with the antenna for nearly a year. But the ubiquitous 3.5° Kelvin signal refused to go away.

The problem was solved, and Penzias and Wilson were catapulted to fame by one of the odd, serendipitous connections that sometimes catalyze a scientific discovery. On a return flight from Puerto Rico, a radioastronomer named Bernard Burke chatted with a colleague named Ken Turner. Turner told Burke about a talk he'd heard by a Princeton theoretician named James Peebles. Peebles, at the suggestion of a senior Princeton researcher, Robert Dicke, had calculated what kind of radiation might be left over from a primeval, universe-creating fireball. Peebles had estimated that the radiation from such a fireball might be detectable as an omnidirectional microwave hum of less than 10° kelvin. Penzias, who knew Burke from a meeting he had attended, called him about an unrelated topic. Burke asked Penzias how things were going. Fine, Penzias said, except for this noise. . . .

Robert W. Wilson (left) and Arno Penzias.

At Burke's urging, and with some trepidation, Penzias called Dicke at Princeton. Dicke happened to be having a brown-bag lunch with Peebles, Peter Roll, and David Wilkinson. Dicke had asked Wilkinson and Roll to build a receiver to listen for the background radiation. After the call from Penzias, Dicke turned to the group to tell them they'd been scooped. A few days later, on Friday March 26, 1965, Dicke, Wilkinson, and Roll visited Penzias and Wilson. After a brief talk, and a tour of the antenna atop Crawford Hill, Dicke was convinced that Penzias and Wilson had found the fossil radiation that he and his team were searching for. Penzias and Wilson were pleased to learn of a possible, and potentially important, source for their "excess noise." But they were cautious about Dicke's cosmological theories. "They seemed to think that the cosmic background radiation was a pretty wacky idea," says Wilkinson. Their caution was misplaced. When the discovery was announced in simultaneous papers by Penzias and Wilson and by Dicke, Peebles, Roll, and Wilkinson, the field of cosmology was transformed.

For decades, two competing cosmological theories had vied for acceptance without adequate data on either side. In 1927, a Belgian priest-physicist, Georges Lemaitre, had used Einstein's relativistic equations to derive an expanding universe whose birth could be traced back to a single "primeval atom." Einstein, who believed that the universe was stable, hated the idea. At a meeting, he told Lemaitre, "Your calculations are correct, but your physical insight is abominable." But in 1929, Edwin Hubble found that distant galaxies are all moving away from each other. Perhaps Lemaitre was right and Einstein was wrong.

In the late 1940s, George Gamow and his associates used an early big bang model to try to calculate the proportions of the chemical elements, that they believed had been cooked up in the primordial fireball. In fact, two of Gamow's colleagues, Ralph Alpher and Robert Herman, had "scooped" Dicke and Peebles by nearly twenty years. In 1948, they published a paper in the prestigious journal *Nature* in which they calculated that the faded echo of the Big Bang must still exist as background radiation at 5° Kelvin.

Unfortunately for Gamow, Alpher, and Herman, their work was swept aside and largely forgotten when Fred Hoyle and other astrophysicists proved that all elements heavier than helium must have formed inside stars and supernovae, not in what Hoyle derisively called a Big Bang. Many cosmologists came to support the Steady-State universe proposed and avidly argued for by Hoyle. According to Hoyle, the universe looked the same at any point in time. New matter popped into existence an atom at a time at just the rate needed to maintain the universe as we see it now. At the time, nobody knew how to test either theory. A distinguished physicist, W. A. Fowler, captured the feeling of many scientists: "Cosmology is mostly a dream of zealots who would oversimplify at the expense of deep understanding," he wrote.

That all changed when Penzias and Wilson discovered the cosmic microwave background radiation. At a stroke, cosmology became a science based on observation and measurement. The echo of the Big Bang had been found; the Steady-State universe was dead. Suddenly the cosmos had a birthday, and science had a stethoscope to listen to its first cry. Penzias and Wilson's discovery triggered an avalanche of research that continues to the present.

We now know that the cosmic background radiation provides a snapshot of the universe when it was less than 300,000 years old. That's when the expanding primordial plasma had cooled enough to allow electrons and nuclei bond into atoms, allowing radiation to flow freely for the first time. Decades of measurements by teams all over the world made it clear that the background radiation is incredibly smooth, varying in temperature from place to place by less than one part in 10,000. That amazing uniformity forced a revision of the Big Bang model. In 1981, Alan Guth explained that remarkable smoothness by proposing inflation—an incredibly brief period during the first fraction of the first second of the existence of the universe, during which it expanded by a factor of 10 to the 50th power, from smaller than the nucleus of an atom to the size of a tennis ball.

Within the past decade, NASA's COBE satellite uncovered the next secret hidden in the cosmic microwave background. In April 1992, a team of researchers revealed long-expected "wrinkles" in the cosmic background—minute differences in temperature that reflect equally minute density variations in the early universe. Theorists breathed a sigh of relief—they could now trace today's galaxies and galaxy clusters to those early irregularities.

The final chapter of the book of the universe has not yet been written. Within the past few years, measurements of supernovae in distant galaxies have shown that the rate at which the universe is expanding is speeding up, not slowing down as the standard Big Bang theory predicts. And, using exquisitely precise new instruments to measure the microwave background, scientists have now found a long-predicted second level of variation reflecting the interaction of matter and energy in the early universe. Both the supernova findings and these "acoustic oscillations" imply that the universe is pervaded by a mysterious "dark energy" which is pushing space apart. But, to date, none of the existing theories of matter and energy can convincingly account for this mysterious repulsive force.

Some 2,600 years ago, Thales asked how the universe came to be. Scientists have made enormous progress toward an answer, much of it in the last fifty years. Within the next decade, astronomers using powerful new instruments and techniques will multiply what we know about the birth and evolution of the universe. Their findings will give theorists yet more information to work with. We are peering ever more deeply into the dawn of time. But nature has yet to reveal its deepest secrets.

33

We Are Not
What We Seem

Human and animal consciousness, as well as other types of biological beauty
and complexity, are properties of our coevolving, pointillist bacterial ances-
try. Cellular interliving, an infiltration and assimilation far more profound
than any aspect of human sexuality, produced everything from spring-green
blooms and warm, wet mammalian bodies to the Earth's global nexus.

—Lynn Margulis, 1998

I don't ever believe authorities.

—Lynn Margulis, 2001

Lynn Margulis (1938–), more than any scientist since Darwin, has
forced a radical revision of evolutionary innovation. Darwin perceived a great tree of life, with each species tracing its origins back to a
single source. His vision unified all life and revealed mankind as simply one branch among many. Genetics, now understood in molecular
detail, promises to allow every twig to be traced back, branch by
branch, to its root. But if Margulis is right, nature has been far more
promiscuous and far more creative than Darwin dreamed. She writes,
"In reality, the Tree of Life often grows in on itself. Species come together, fuse, and make new beings, who start again. . . . The tree of
life is a twisted, tangled, pulsing entity with roots and branches meeting underground and in midair to form eccentric new fruits and hybrids." Those eccentric hybrids, Margulis has shown, include all
organisms more complex than bacteria, including humans.

Margulis showed signs of her talents at an early age. She was a precocious child and gifted student, entering the University of Chicago when she was just fourteen. It was there that she met an equally bright nineteen-year-old graduate student, Carl Sagan, whom she later married. She credits the University of Chicago, under the innovative Robert M. Hutchins, for nurturing her lifelong critical skepticism. Margulis studied genetics and evolution at the University of Wisconsin, Madison, and at the University of California, Berkeley. She taught at Boston University for twenty-two years, and is now a notoriously energetic and communicative Distinguished Professor at the University of Massachusetts, Amherst. From the start, she found herself fascinated by genetic material outside the cell nucleus, then called cytoplasmic genes. How did they come to be there, she asked, and what do they do? Her skepticism flared when she read the geneticist Thomas Hunt Morgan's famous dictum, "From the point of view of heredity, the cytoplasm of a cell can safely be ignored." Her studies soon led her to the opposite conclusion. "It seemed obvious to me that there were double inheritance systems with cells inside cells." Much of her career can be seen as an attempt to make this "obvious" truth obvious to other biologists.

Since Darwin's time, the field of evolution has struggled with the problem of variation. Variation provides the raw material, the experiments, the trial balloons, which natural selection discards or keeps. In the decades after Darwin's death, first mutation and later recombination—both of which rewrite sections of the genetic text—emerged as sources of variety edited by natural selection. Margulis, however, believes that nature's creativity is fueled by a far more powerful source—symbiosis. The classic examples of symbiosis—different organisms living in intimate, interdependent association—are lichens (composed of algae and fungi), cellulose-digesting bacteria in the guts of termites, and nitrogen-collecting bacteria in the roots of many plants. But Margulis sees these relationships as recent and partial. She believes that separate branches of the Tree of Life cross and merge in a process called symbiogenesis. Margulis defines symbiogenesis as evolutionary change by the inheritance of acquired sets of genes. It's the ultimate in cooperation, and it paints a very different picture than Darwin's view of incessant competition as the engine of evolution. Symbiogenesis results in creatures that are radically different and much more complex than either "parent." Natural selection, Margulis argues, chooses new life forms from a pool of applicants that is far more varied than earlier biologists dreamed. "What we've saying is that the way you pass from [one] species to another species is the inheritance of acquired whole genomes," says Margulis. "The acquisition not of something random, but something like a bacterium that's already highly [evolved]."

This provocative and disturbing idea turns out to have deep roots. Almost as soon as symbiosis was discovered, a few scientists suggested that it might help drive evolution. In the nineteenth century, Andreas Schimper proposed that chloroplasts, the solar energy organs within plant cells, might be the product of symbiosis. Another German biologist, Richard Altmann, argued a similar case for mitochondria, the membrane-bound energy factories powering animal cells. Similar ideas were developed by the Russians Andrei Famintsyn and Konstantine Merezhkovskii, and by the Americans Ivan Wallin and Albert Schneider. It was Merezhkovskii who, in 1910, coined the term "symbiogenesis." But for nearly a century mainstream biologists refused to take such unseemly and unauthorized unions seriously.

In 1967, Margulis first published her step-by-step outline of the origin of eukaryotic cells, the complex, nucleated cells that we share with all organisms except bacteria. Her radical paper, "Origins of Mitosing Cells," was rejected nearly a dozen times before appearing in the *Journal of Theoretical Biology*. Her ideas have since evolved into the serial endosymbiosis theory, which details how the cells of plants, animals, fungi, and other organisms evolved through a specific series of symbiotic mergers between different types of organisms early in life's history. It was these dramatic innovations, Margulis argues, not gradual Darwinian evolution, that transformed ancient bacteria into the complex, multiorgan cells we are made of. Margulis may have solved what the pioneering scientist, J. D. Bernal, had identified as one of biology's greatest mysteries—the origin of nucleated cells.

Margulis finds traces of four vital mergers between different kinds of early bacteria. The first union was between a heat-loving archaebacterium and a swimming bacterium, perhaps a spirochete. This produced a more complex swimming microorganism that would eventually evolve into cells sequestering most of their genetic material within a membrane-bound nucleus. According to Margulis, the genes that initially coded for wriggling, spirochete-like appendages were co-opted over millions of years to produce the organizing centers and filaments that pull genetic material to opposite sides of a cell before it reproduces by splitting in two. In her view, this gene-handling device,

Lynn Margulis.

stemming from symbiogenesis, is what distinguishes all complex life forms from bacteria.

This new kind of creature then engulfed an oxygen-burning bacterium, which evolved over time into the mitochondria now found in the cells of all organisms with nucleated cells. And finally, this complex, swimming, oxygen-processing, one-celled organism engulfed a photosynthesizing bacterium destined to become the light-processing chloroplasts and related plastids of today's algae and plants. Margulis believes that these multiple bacterial unions, which created the first mobile, oxygen-breathing, nucleated cells, took place more than 2 billion years ago.

Although initially ignored or scorned by most biologists, the serial endosymbiosis theory has finally won acceptance—it even appears in high school textbooks. The case for the origin of mitochondria as ancient, oxygen-processing bacteria has been settled, along with cyanobacteria, as the source of chloroplasts in plant cells. Chloroplasts and mitochondria are now known to carry their own DNA, whose molecular code can be traced back to specific kinds of bacteria, just as Margulis predicted. Electron microscopy has shown that the tails of sperm, the cilia of mobile microbes, and the cilia in human throats and fallopian tubes all share the same structure—eleven sets of microtubules in a particular nine-plus-two arrangement. Still, Margulis's claim that wriggly spirochetes contributed the genetic know-how to form the cilia and flagella that give cells internal and external mobility remains controversial. Margulis and her students are continuing to carry out research seeking to prove this final aspect of their case. But the few remaining genetic and physiological clues from this ancient union are subtle and subject to conflicting interpretations.

Spurred by her new view of life, Margulis and her colleague Karlene Schwartz have campaigned to change how all living things are classified. They have discarded the ancient view that all species can be sorted into plants and animals, the nineteenth-century division into animals, plants, and microbes, and the more recent bifurcation into prokaryotes (cells without nuclei) and eukaryotes (one-celled or multicellular organisms with cell nuclei). Margulis, building on work by Robert H. Whittacker, splits all life into five kingdoms, which she compares to a human hand. Bacteria, or Monera, form the base of the hand and thumb. They are separated from all other living things because their cells do not have nuclei, *but also by virtue of not having formed through symbiosis.* Each finger represents a branch of life ultimately derived from bacteria—*not just from one kind, but by the symbiotic union of several ancient bacterial progenitors.* The proof of that ancient merging, Margulis believes, is the cell nucleus itself, shared by all non-bacterial species.

Three of the "fingers" are familiar: animals, plants, and fungi (mushrooms, yeasts, and molds). The fifth kingdom, called Protoctista, contains a strange melange of organisms, including most of the "animalcules" Leeuwenhoek discovered and described so vividly. It includes algae and kelp, formerly considered primitive plants, and amoebas, earlier lumped with animals. It's home to some very common creatures few people think about, such as the microorganisms that form poisonous red tides, and the diatoms and coccoliths whose minute shells form vast deposits, including the White Cliffs of Dover. It also contains some truly weird forms of life, like the slime molds that spend most of their lives as separate cells scattered in the soil, but which come together to form a complex multicellular creature that reproduces by making and broadcasting spores.

Margulis has at least one strong competitor challenging her five-kingdom view of life. The microbiologist Carl Woese, based on differences in the RNA molecules organisms use to build proteins, sees just three great "domains" of life, Eukarya (organisms whose cells have nuclei), Eubacteria (true bacteria), and Archaea. Archaea are one-celled organisms without nuclei, many of which thrive in hot springs, acidic solutions, brine, or other environments that seem hostile to life, but which may reflect conditions at the dawn of life on Earth. Margulis does not think the Archaea and true bacteria differ as profoundly as both do from organisms whose cells have nuclei.

The serial endosymbiosis theory has spurred a generation of scientists to look for symbiotic relationships. Their work has shown that symbiosis is a fundamental natural theme. Nine out of ten kinds of land plants depend on fungi in their roots to get crucial nutrients from the soil. Almost all plant-eating insects and animals—including the animals we depend on for food—would starve without the help of microbes in their guts with the chemical know-how to digest cellulose. And researchers are discovering species in the process of merging. Kwang Jeon, a zoologist at the University of Tennessee, caught bacteria in the act of invading amoebas and creating a new kind of organism. Japanese biologists recently found a bacterium that lives within the cells of aphids. One-tenth of the bacteria's genes make substances the aphid needs but can't make for itself, while the aphid provides the raw materials to make the bacteria's cell membrane. Neither organism can survive or reproduce without the other. This looks like a replay of how mitochondria and other cell organelles formed billions of years ago. Scientists have observed bacteria fusing with and transferring genes to mammalian cells. Molecular biologists have found that our chromosomes contain the remnants of many viruses that managed to worm their way into the human genome. Known as endogenous retroviruses, or ERVs, some of these once-alien genes now appear to play key roles in human development.

Copernicus, Galileo, and Darwin all destroyed some of our most deeply rooted assumptions about our central place in the scheme of things. Margulis adds her own resounding voice to that profound but humbling chorus. "All of the sciences tell us that humans are extremely recent and *weedy*—the weedy mammal," she says. "We have a lot of planet-mates that we tend to denigrate." What Margulis has shown is that our very cells betray the fact that we are far more intimately and intricately related to those planet-mates than anyone before her dreamed.

34

Planetary Pioneers

Do there exist many worlds, or is there but a single world?
This is one of the most noble and exalted questions in the study of Nature.

—*Albertus Magnus (1200–1280)*

It's a very strange feeling, to have something new based only on the data,
something completely unexpected. I can tell you, it's very scary.

—*Didier Queloz*

What excites me about the exoplanets is that
everything is yet to be discovered.

—*Michel Mayor*

Until 1995, astronomers speculating about the birth and development of planets had just one planetary system to study—our own. Since then, observers have found nearly one hundred planets, most orbiting stars like the Sun. This burgeoning collection is full of surprises—Jupiter-sized planets whipping around their stars in sub-Mercury-sized orbits, other gas giants careening toward and away from their stars like loose cannonballs, and even lonely planets without stars. We now know of a half-dozen sunlike stars with full-fledged planetary families, and of a similar number of planets (all much larger than Earth) circling their stars in cozy orbits where it's neither too hot nor too cold to support life. Although astronomers had been diligently trolling for planets around distant stars for more than fifty years, this flood of discovery was started by two astronomers who had just begun their search, and who mainly hoped to find dwarf stars.

The planet quest has ancient roots. Anaximander, in the sixth century B.C., was the first to speculate that the Earth was not unique. The atomists who followed him argued that an infinite number of atoms following natural laws must form many worlds. The Roman poet Lucretius echoed their beliefs, and, a millennium later, so did the fervent Giordano Bruno. For this and other heresies, the Inquisition condemned him, and he was burned at the stake in Rome in 1600. In the middle of the eigtheenth century, the Prussian philosopher-scientist Immanuel Kant pieced together his "nebular hypothesis," a qualitative but surprisingly accurate description of how a star and its retinue of planets could form from the gravitational collapse of a cloud of gas and dust into a rotating disk. During the twentieth century, theorists refined his theory into a convincing picture of how our own solar system formed, with temperature differences in the disk leading to small, rocky planets circling close to the Sun, giant, gas-shrouded planets farther out, and icy debris—potential comets—lurking at the fringes.

But with just one solar system to study, nobody knew how accurate the model might be. How many stars formed planets? How many planets survived in stable orbits? Was Earth unique, or were there many planets where life could exist? Until 1995 we knew little more than the ancient Greeks.

Throughout the second half of the twentieth century, many astronomers searched for planets around nearby stars. They knew that they could not hope to see even a planet as large as Jupiter; its faint light would be swamped by the glare of its star. But they could hope to detect a giant planet through its gravitational tug on its star. A planet and its star revolve around their common center of gravity, with the planet tracing a large orbit, its star a tiny one. Some observers thought they might be able to detect a star's side-to-side wobble by carefully measuring its position in the sky over many years. Others hoped to capitalize on the Doppler effect by detecting minute changes in the color of a star's light caused by its motion toward or away from Earth. By 1995, several teams of astronomers had refined spectroscopic techniques to the point that they could measure the movement of a distant star to an accuracy of thirteen meters per second—not much faster than an Olympic sprinter. The Swiss astronomers Mayor and Queloz were the most recent members of this club.

Michel Mayor (left) and Didier Queloz (right).

Like many astronomers, Michel Mayor (1942–) had been fascinated by the stars since childhood. "The Scouts had different specialties, like cooking and topography," he says. "I had the specialty of 'astronomer.'" As a graduate student, he studied how galaxies form, and then spent many years refining spectroscopic techniques to search for brown dwarfs—cool, dim objects that are thought to form like stars but which fail to grow massive enough to support hydrogen fusion. Like most scientists, Mayor is passionate about his work. But he's also a family man and something of a romantic. He lovingly describes his wife, Francoise, as his scientific muse.

Following the death of his long-term collaborator, Antoine Duquennoy, Mayor began to work with a young graduate student, Didier Queloz

(1966 –). Queloz had been interested in science as a child, studied physics and mathematics, and had decided to specialize in astrophysics. "I liked nature, visiting remote countries, and mountains," Queloz recalls. "I thought that astrophysics would allow me to do that." Queloz describes Mayor as an ideal mentor. "Our relationship was just great," he says. "He's very friendly, and he always trusts me. Only once, he asked me, 'Are you sure of the data?' I said yes. That was it."

In April 1994, observing at the Haute-Provence Observatory in southern France, Mayor and Queloz started to study 142 nearby, sun-like stars with the help of a cleverly designed new spectrometer. They were looking for telltale shifts in a star's spectrum that could tip them off to the presence of a "low-mass stellar companion." If they were lucky, they thought, they might find some brown dwarfs, adding to the very short list of such failed stars. Since the thirteen-meter-per-second sensitivity of their instrument was just about equal to Jupiter's tug on the Sun, they knew they had little chance of finding a planet.

In early January 1995, one star—the fifty-first brightest in the constellation Pegasus—caught Queloz's eye. The star showed a dramatic wobble, well beyond any possible errors in their measurements. Queloz knew he had found something, but he did not know what. He concentrated on the star, at one point observing it every night for an entire week. After checking and rechecking the instruments, and making sure that 51 Peg was a mature, stable star, Queloz calculated the object's orbit and faxed it to Mayor in Hawaii. "I was a little naive," says Queloz. "Because I didn't have a lot of background, I believed it could be real, could be a planet." The star was not visible between March and June, but during that time they refined their calculations of the object's orbit.

When 51 Peg could be observed again early in July, Mayor and Queloz had predicted just what velocity their spectroscope should reveal. They must have been quite sure of themselves, since they brought their children to the observatory and came armed with champagne and cake. Although Mayor points out that the discovery took place over many months, when their observations exactly matched their prediction they allowed themselves a moment of celebration. "It was like a time clock," says Queloz, "which you expect with a planet." Queloz adds that his two-year-old did not understand what they had accomplished but did enjoy the cake.

The proud fathers had found an extra-solar planet, the first ever detected around a sunlike star. They had provided an answer to a 2,500-

year-old question, proving that our solar system is not unique. But their discovery instantly raised new questions. The object they had discovered was unlike anything in our solar system. They could estimate its mass, so they knew it was a giant planet, probably a gas giant like Jupiter or Saturn. But instead of orbiting far from its star, as the giant planets of our solar system do, and as predicted by theory, this planet skimmed its star at one-twentieth the distance from Earth to the Sun. Each orbit took just 4.2 days. In comparison, Mercury—the Sun's closest and speediest companion—orbits in a leisurely eighty-eight days. Mayor and Queloz were not even sure that a planet could survive so close to a star.

The fifty-year hunt for extra-solar planets had a long history of mistaken claims. Mayor and Queloz did not want to join the list of astronomers who had humiliated themselves by announcing the discovery of a planet, only to find they had made some error. They spent the next months ruling out every other possibility. Perhaps 51 Peg was a variable star, puffing up and contracting on a 4.2-day schedule. That could account for their Doppler measurements. But such stars also change in brightness. They found that 51 Peg's brightness stayed perfectly steady. Perhaps the star had huge sunspots, although it was a bit old for such problems. If it happened to rotate every 4.2 days, the moving sunspots could fool them. But they could estimate the star's rotation from the degree to which the star's spectral lines were blurred. It rotated about once a month, much like the Sun. Perhaps the star and its companion happened to be revolving at nearly a right angle to our line of sight. In that case their spectroscope would detect only a fraction of the object's motion, leading them to underestimate its mass. It might be a dwarf star, not a planet. But in that case they would also not see any spreading of the star's spectral lines. By October they were sure enough of their discovery to share it with the world.

Mayor and Queloz presented their strange new planet at an astronomical conference in Florence, Italy, on October 6, 1995. It was a fitting location, not far from the villa where Galileo, imprisoned by the Inquisition for arguing that the Earth is not the center of the universe, spent his last days. Mayor and Queloz gave the world not just the first planet around a distant, sunlike star, but a new kind of planet, a gas giant sizzling at 1,000° Celsius because of its incredible closeness to its star. While a few found this new kind of planet too strange to be believed, many of the leading astronomers at the meeting were as excited as Mayor and Queloz.

Word spread quickly to Geoff Marcy and Paul Butler, two astronomers in California who, until then, considered themselves the
front-runners in the race to find the first extrasolar planet. They had
been making precise spectrographic observations of more than 100
nearby stars for nearly a decade. With enormous diligence, they had
refined their equipment and calculations to the point that they could
measure stellar velocities to an astonishing three meters per second—
jogging speed. They had compiled thousands of observations and accumulated gigabytes of data. Unfortunately, they had not analyzed
most of their data. The research grants they had managed to get could
not pay for the large amounts of computer time needed to turn their
data into usable form. And they had assumed, as had almost every
other observer and theorist, that any planet massive enough for them
to detect would have an orbit measured in years. So it made sense to
pile up years of observations before analyzing them. Their strategy had
seemed sensible. But, like Rosalind Franklin studying DNA, they'd
been scooped.

Marcy and Butler, however, did not withdraw from the race. They
happened to have four nights of observing time available. Within a
week they confirmed Queloz and Mayor's discovery. They then pulled
every possible string to get computers and time to analyze their enormous backlog of data. Before the end of 1996 they had mined their
data and found a half dozen new planets.

Since October 1995, Mayor and Queloz, Marcy and Butler, and
an increasing number of other planet-hunting teams have discovered
close to one hundred exoplanets, with far more to come. It's a large
enough number to give theorists their first-ever chance to test competing theories about how planets form and evolve. Several classes of
planets have emerged. These include a large number of "hot Jupiters,"
of which 51 Peg is typical. The theorist Douglas Lin was among the
first to point out that these gas giants cannot have formed so close to
their stars. He has shown that tidal interactions within a planet-
forming disk doom most planets to spiral in toward their stars. Some,
like the planet orbiting 51 Peg, stop their death spiral just in time.
Many others, he believes, plunge into their stars.

In our solar system, all the planets except tiny Mercury and distant
Pluto orbit the Sun on essentially circular paths. That's not the rule
elsewhere. Nearly a third of the planets found so far zoom toward and
away from their stars in stretched-out, cometlike ovals. Scientists who
would like to believe that habitable, Earth-like planets are common

have had to admit that both migrating planets and those on cometlike orbits would destroy any small, rocky planets that might have formed. However, they are encouraged by the sheer number of planets being found, and by the fact that at least some are in stable orbits within their stars' habitable zones.

Current techniques will uncover more giant planets, but they are not capable of finding lightweight planets like Earth. An approach that may provide the first sign of an Earth-sized exoplanet is gravitational microlensing—based on Einstein's discovery that the gravitational field of a massive object bends light. If a star and its planets happen to pass between Earth and a background star, the light from the more distant star will brighten and dim in a predictable way. Even a planet as small as Mars could be caught this way. Such events are rare, but astronomers are now able to monitor large areas of the sky at once to watch for such events. Queloz, however, is betting that the first detection of an Earth-sized planet will be accomplished by a different method—measuring the slight dimming of a star's light as a planet passes across the face of the star as seen from Earth. He expects this transit approach to turn up an Earth-like planet by 2010.

Astronomers, of course, want to be able to see and study exoplanets directly. Optical or infrared interferometry appears to be the key. By combining the light from widely separated telescopes, it's possible to resolve much finer detail than with an ordinary telescope. In addition, an interferometer can be adjusted so that the light from a star is cancelled out while light from its planets is enhanced. Observers are just starting to use large Earth-based interferometers, and may soon be able to detect Jupiter-sized planets directly. Within two decades, astronomers hope to have huge interferometers in space. The second generation of these instruments, in orbit around the Sun far from Earth, should be able to let us see and study Earth-like planets for the first time. Spectroscopy will again prove vital, since it will allow scientists to detect oxygen, ozone, and other signs of life.

Mayor and Queloz showed us that the ancient atomists were right—nature is capable of making many planets. We now know that planets and planetary systems come in bewildering variety. But we still do not know if, among all the kinds of worlds nature has made, any besides Earth are warm and wet, potential homes for living things. However, it may not be too long before an array of telescopes, orbiting in deep space, relays back to Earth the blue-tinged signature of a second living planet.

35

After Dolly, Life Will Never Be the Same

[T]he cloning of mammals by simple nuclear transfer
is biologically impossible.

—*Davor Solter and James McGrath*, Science, *December, 1984*

One advantage I had is that I don't believe
what people tell me. I never did.

—*Keith Campbell*

Even if they did not grasp [Dolly's] full significance . . . people felt that life
would never be quite the same again. And in this they are quite right.

—*Ian Wilmut*

It basically means there are no limits.
It means all of science fiction is true.

—*Lee Silver, Princeton University*

Ian Wilmut (1944–) is an unlikely pioneer. He's not an iconoclast, he's not driven to be first, and he's not after fame or fortune. He's a balding, soft spoken Englishman, married, with three grown children, whose most distinctive features are his reddish beard and clear blue eyes. His approach to science, by his own description, is "logical and methodical." Wilmut set out to be a farmer, and only switched to science when he realized that farming required more business savvy than he thought he had. His collaborator, Keith Campbell (1954–), is a bit more willing to take risks—either on his mountain bike or in pursuit of an exciting idea. And he's never been slowed down by being told that what he's trying to do is impossible. It may be that the balance between bold, bright ideas and cautious, step-by-step testing is what enabled Wilmut and Campbell to create Dolly the sheep. Dolly was the first animal cloned from the cell of an adult. Her birth shocked the scientific world and made Wilmut and Campbell famous. Today, they and hundreds of researchers around the world are avidly exploring, colonizing, and exploiting the brave new world Dolly opened up. That world will almost certainly include human clones, most likely surprisingly soon.

Dolly was born in the late afternoon of July 5, 1996, on the grounds of the Roslin Institute a few miles from Edinburgh, Scotland. She weighed in at a hefty 6.6 kilograms, or 14.5 pounds, but otherwise appeared to be a typical Finn-Dorset lamb, with soft, grayish-white fleece and a white face. Her background of course made her anything but typical. She bore no relation to her surrogate mother. In fact, she did not have either a mother or a father. In a miniature act of creation, complete with a tiny electrical jolt, Wilmut and Campbell had formed her by fusing a mammary cell from an anonymous, long-dead ewe with an unfertilized egg stripped of its chromosomes. With her birth, religious leaders, ethicists, politicians, and millions of ordinary people came face to face with the possibility of human cloning. Scientists were astonished for other reasons. Twenty-five years of cloning attempts had convinced most of them that cells from anything other than a very early embryo could not be cloned. Once cells had taken a step or two toward their adult roles, biologists believed, they could not go back. In addition, biologists were convinced that the genes themselves, not anything in the cytoplasm, controlled the genetic program. Dolly broke all the rules.

Perhaps ten years of hard work had taken their toll. Or maybe Wilmut and Campbell were not fully aware of the firestorm that Dolly's birth would ignite. For them her arrival passed nearly unnoticed. Campbell was on vacation with his wife and children. Wilmut had set aside a bottle of champagne, but realized that it didn't feel

Ian Wilmut and Polly.

right to drink it without his colleague. He doesn't actually remember receiving the telephone call telling him that Dolly was alive and well, or what he did when he heard the news. Wilmut is pretty sure, however, that he took his dog on its usual walk in the hills.

Although cloning pushed them into the limelight, Wilmut and Campbell never viewed it as anything more than a stepping-stone to other scientific and practical goals. Their scientific aim—and the force driving the decades of cloning research that preceded them—was to understand how an organism develops from a single cell to an adult animal. That remarkable, still-mysterious journey has been described as "the central problem of biology." Struggling with this problem, molecular biologists had come to realize that an organism's genome is not just a grab bag of instructions for making proteins. It's more like a computer program or a musical score. In the right environment—the cytoplasm of an egg cell—the genetic program runs in a stunningly complex, but highly orchestrated sequence. The genetic program is a useful metaphor, but in biology it's the messy details that count.

In addition to their scientific aims, Wilmut and Campbell had some very practical goals. The Roslin Institute, where they worked, and its spin-off, PPL Therapeutics, wanted to engineer animals that would produce valuable chemicals, such as drugs to treat human diseases. Trudging toward that goal, Wilmut spent years trying to slip desirable genes into embryonic cells one cell at a time. Then, early in 1987, over beer in a Dublin pub, he heard a rumor that the highly respected embryologist Steen Willadsen had cloned a sheep, using a kind of embryonic cell that had already started to differentiate. Unlike the earliest embryonic cells, these were plentiful. "I felt pennies dropping and bells clanging," Wilmut recalls. If he could add genes to hundreds or thousands of such cells at once, he could select the best, fuse them with egg cells, and produce flocks of drug-producing sheep. His problems would be solved. It was a moment of inspiration, but one that would take a decade to realize.

When Campbell and Wilmut started toward that goal, most biologists believed that the developmental program could only run in one direction. A fertilized egg splits into two smooth, round cells. Those cells divide again and again, eventually differentiating into all the distinct kinds of cells an organism needs—skin, gut, heart, nerves, and dozens more. Scientists believed that once cells had committed themselves to a particular fate there was no going back. Campbell, however, had studied cancer cells. Although tumors typically start from a single cell, Campbell knew that they often turn out to contain many different kinds of cells. Under certain circumstances the genetic program could be reset.

Campbell made a second key discovery on the road to Dolly. After several years of preliminary work, he came to suspect that a particular stage of a cell's normal cycle might put the nucleus into a receptive state, one in which its DNA could be reached and "reprogrammed" by substances in the cytoplasm of an egg. He knew that the egg cell's cytoplasm orchestrates the first several days of embryonic life, while the embryo's genes remain inactive. He came to believe that quiescent cells—those that were not busy transcribing genetic information into proteins—were more likely to be receptive to those cytoplasmic factors. If he was right, differentiated quiescent cells, maybe even cells

Keith Campbell.

from an adult, could be reprogrammed. The experiment that produced Dolly from a fully differentiated adult cell was designed to test these ideas.

Dolly's birth in 1996 seemed to support Campbell's intuition. Like a dip in the river Nepenthe, immersion in the cytoplasm of an egg cell had reprogrammed the quiescent donor cell nucleus. However, in May 1998, a research team led by Jose Cibelli at the University of Massachusetts, Amherst, produced three cloned calves from non-quiesecent cultured fetal cells. And in late 1999, researchers at the University of Hawaii cloned mice from metabolically active embryonic stem cells. Campbell now believes that several different stages in the cell cycle may support reprogramming. He and many other scientists are now trying to understand developmental reprogramming at the molecular level. Dolly proved that it can occur, but nobody knows just how it takes place.

Dolly and Bonnie.

Although Wilmut and Campbell can list dozens of important scientific benefits of the technology they created, it's the potential for human cloning that has stirred up a storm of controversy. Human cloning seems to have touched a raw nerve. A poll taken soon after Dolly's birth found that ninety-three percent of Americans thought human cloning should be illegal. President Clinton agreed, and immediately banned the use of federal funds for research in that area. Thirteen European countries have banned human cloning research. Surprisingly, Wilmut and Campbell also strongly oppose human cloning and related research. "Any kind of manipulation of human embryos should be prohibited," says Wilmut.

Wilmut points out that cloning by nuclear transfer is still extremely inefficient. Although researchers are rapidly improving the odds, typically hundreds of cells must be fused and nurtured to produce one healthy cloned animal. The rates of failed pregnancies, stillbirths, and deaths soon after birth are higher—often far higher—than in normal reproduction. With more animals being cloned, a new condition has emerged—the large fetus syndrome. Cloned animals are often as much as a third larger and heavier than normal at birth, placing their surrogate mothers and themselves at risk. Those that survive sometimes become enormously obese, or show other developmental and metabolic problems. Dolly herself has developed arthritis at an early age. Wilmut suspects that reprogramming does not restart the genetic program uniformly and completely. As a result, the program does not run smoothly, producing a gamut of developmental problems. He simply does not want human fetuses subjected to such risks. At the same time, he does not want legislation aimed at preventing human cloning to block research into legitimate medical and scientific uses of the technology.

Dolly's birth seems to have jolted biotechnological research with animals into high gear. Within months of her birth, Wilmut and Campbell cloned another sheep, Polly. She carries genes that allow her to secrete Human Factor IX, a costly blood-clotting drug used to treat hemophilia, in her milk. PPL Therapeutics now milks a herd of several hundred medication-producing sheep. Other researchers have cloned cattle, mice, and even rhesus monkeys. Biologists are now able to add genetic switches to the genes they insert, allowing them to turn them on and off at will. Cloned animals will soon be producing a host of pharmaceuticals as well as growing tissues and organs for transplantation into humans. Other animals may be cloned to help preserve endangered species or to create herds of disease-resistant, high-producing farm animals.

All of these technologies, plus unexpected new ones, will proliferate from the germinal ideas that led to Dolly. Along with the potential for human cloning, they place unprecedented power in human hands. In the first half of the twentieth century, physicists unlocked the secrets of the nucleus of the atom, giving mankind unprecedented power to create or destroy. In the last years of the same century, biologists unlocked the potentials of the nucleus of the cell. We now have the capacity to change and potentially even direct our own biological development and destiny—as individuals and as a species. The genie truly is out of the bottle. What remains to be seen is where our wishes and desires will take us.

References and Further Reading

Introduction

Bronowski, Jacob. *The Ascent of Man*. Boston: Little Brown and Company, 1973.

Huff, Toby E. *The Rise of Early Modern Science: Islam, China, and the West*. Cambridge: Cambridge University Press, 1993.

Needham, Joseph. *Science and Civilisation in China*. Cambridge: Cambridge University Press, 1954.

Neugebauer, Otto. *The Exact Sciences in Antiquity*. Providence, R.I.: Brown University Press, 1957.

Palter, Robert M., ed. *Toward Modern Science*. Vol. I. New York: Farrar, Strauss & Cudahy, 1961.

Ronan, Colin. *The Astronomers*. London: Evans Brothers, 1964.

Ross, Frank. *Oracle Bones, Stars, and Wheelbarrows: Ancient Chinese Science and Technology*. Boston: Houghton Mifflin, 1982.

Temple, Robert. *The Genius of China*. New York: Simon and Schuster, 1986.

1. Thales and Natural Causation

Farrington, Benjamin. *Greek Science: Its Meaning for Us*. London: Penguin, 1953.

Heath, Thomas. *A History of Greek Mathematics*. Vol. I, *From Thales to Euclid*. Oxford: Oxford University Press, 1921.

Lloyd, G. E. R. *Early Greek Science: Thales to Aristotle*. New York: Norton, 1970.

Long, A. A., ed. *The Cambridge Companion to Early Greek Philosophy*. Cambridge: Cambridge University Press, 1999.

Neugebauer, Otto. *The Exact Sciences in Antiquity*. Providence, R.I.: Brown University Press, 1957.

West, M. L. *Early Greek Philosophy and the Orient*. Oxford: Clarendon Press, 1971.

2. Anaximander Orders the Cosmos

Guthrie, W. K. C. *The Greek Philosophers: From Thales to Aristotle*. New York: Harper & Row, 1950.

Heidel, William A. *The Frame of the Ancient Greek Maps*. New York: Arno Press, 1976.

Kahn, Charles H. *Anaximander and the Origins of Greek Cosmology*. New York: Columbia University Press, 1960.

Kirk, G. S., and J. E. Raven. *The Presocratic Philosophers*, 2d ed. Cambridge: Cambridge University Press, 1988.

Long, A. A., ed. *The Cambridge Companion to Early Greek Philosophy*. Cambridge: Cambridge University Press, 1999.

Thrower, Norman J. W. *Maps and Civilization: Cartography in Culture and Society*. Chicago: University of Chicago Press, 1972.

3. Pythagoras Numbers the Cosmos

Heath, Thomas. *A History of Greek Mathematics, Vol. 1: From Thales to Euclid*. Oxford: Oxford University Press, 1921.

Koestler, Arthur. *The Sleepwalkers: A History of Man's Changing Vision of the Universe*. London: Hutchinson & Co., 1959.

Long, A. A., ed. *The Cambridge Companion to Early Greek Philosophy*. Cambridge: Cambridge University Press, 1999.

Mourelatos, Alexander P. D., ed. *The Pre-Socratics, A Collection of Critical Essays*. New York: Anchor Press, 1974.

Neugebauer, Otto. *The Exact Sciences in Antiquity*. Providence, R.I.: Brown University Press, 1957.

The Pythagorean Sourcebook and Library. Edited by David R. Fiedler from materials compiled and translated by Kenneth S. Guthrie. Grand Rapids, Mich.: Phanes Press, 1987.

4. Atoms and the Void

Allen, R. E., ed. *Greek Philosophy: Thales to Aristotle*. New York: Free Press, 1991.

Bailey, Cyril. *The Greek Atomists and Epicurus: A Study*. New York: Russell & Russell, 1964.

Cartledge, Paul. *Democritus*. New York, Routledge, 1999.

Long, A. A., ed. *The Cambridge Companion to Early Greek Philosophy*. Cambridge: Cambridge University Press, 1999.

Pullman, Bernard. *The Atom in the History of Human Thought*. Oxford: Oxford University Press, 1998.

5. Aristotle and the Birth of Biology

Ackrill, J. L., ed. *A New Aristotle Reader.* Princeton, N.J.: Princeton University Press, 1987.

Aristotle. *Selections.* Edited by W. D. Ross. New York: Scribner's, 1927.

Edel, Abraham. *Aristotle and His Philosophy.* Chapel Hill, N.C.: The University of North Carolina Press, 1982.

Farrington, Benjamin. *Greek Science.* Vol. I, *Thales to Aristotle.* Harmondsworth, England: Penguin Books, 1944.

Moore, John A. *Science as a Way of Knowing: The Foundations of Modern Biology.* Cambridge: Harvard University Press, 1993.

6. Aristarchus, the Forgotten Copernicus

Cajori, Florian. *A History of Physics.* New York: Dover, 1962.

Cohen, I. Bernard. *Revolution in Science.* Cambridge: Harvard University Press, 1985.

Heath, Sir Thomas. *Aristarchus of Samos: The Ancient Copernicus.* New York: Dover, 1981.

Koestler, Arthur. *The Sleepwalkers: A History of Man's Changing Vision of the Universe.* London: Hutchinson & Co., 1959.

Lindberg, David C. *The Beginnings of Western Science.* Chicago: University of Chicago Press, 1992.

Ronan, Colin. *The Astronomers.* London: Evans Brothers, 1964.

7. Archimedes's Physics

Dijksterhuis, E. J. *Archimedes.* Copenhagen: Ejnar Munksgaard, 1956.

Heath, T. L. *The Works of Archimedes.* New York: Dover, 1953.

Stein, Sherman. *Archimedes: What Did He Do Besides Cry Eureka?* Washington, D.C.: The Mathematical Association of America, 1999.

8. Ibn al-Haitham Illuminates Vision

Alhazen. *Opticae Thesaurus.* Edited by David C. Lindberg. New York: Johnson Reprint Corporation, 1972.

Grant, Edward. *Physical Science in the Middle Ages.* New York: John Wiley & Sons, 1971.

Lindberg, David C., ed. *Science in the Middle Ages.* Chicago: University of Chicago Press, 1978.

Nasr, Seyyed Hossein. *Islamic Science: An Illustrated Study*. London: World of Islam Festival Publishing Company, 1976.

Nasr, Seyyed Hossein. *Science and Civilization in Islam*. Cambridge: Harvard University Press, 1968.

Palter, Robert M., ed. *Toward Modern Science*. Vol. I. New York: Farrar, Strauss & Cudahy, 1961.

9. Copernicus Moves the Earth

Cohen, I. Bernard. *Revolution in Science*. Cambridge: Harvard University Press, 1985.

Copernicus, Nicolaus. *De Revolutionibus*. The manuscript can be found on-line at http://www.bj.uj.edu.pl/bjmanus/revol/titlpg_e.html

Crowe, Michael J. *Theories of the World from Antiquity to the Copernican Revolution*. New York: Dover, 1990.

Dreyer, J. L. E. A *History of Astronomy from Thales to Kepler*. New York: Dover, 1953.

Koestler, Arthur. *The Sleepwalkers: A History of Man's Changing Vision of the Universe*. London: Hutchinson & Co., 1959.

Gingerich, Owen, ed. *The Nature of Scientific Discovery: A Symposium Commemorating the 500th Anniversary of the Birth of Nicolaus Copernicus*. Washington, D.C.: Smithsonian Institution Press, 1975.

10. Galileo Discovers the Skies

De Santillana, Giorgio. *The Crime of Galileo*. Chicago: University of Chicago Press, 1955.

Finocchiaro, Maurice A. *Galileo on the World Systems: A New Abridged Translation and Guide*. Berkeley: University of California Press, 1997.

Galilei, Galileo. *On Motion and On Mechanics*. Translated by Stillman Drake. Madison, Wis.: The University of Wisconsin Press, 1960.

MacLachlan, James. *Galileo Galilei: First Physicist*. New York: Oxford University Press, 1997.

Ronan, Colin. *The Astronomers*. London: Evans Brothers, 1964.

Singer, Charles. A *Short History of Scientific Ideas to 1900*. New York: Oxford University Press, 1959.

Sobel, Dava. *Galileo's Daughter*. New York: Walker & Company, 1999.

MacHamer, Oeter K., ed. *The Cambridge Companion to Galileo*. Cambridge: Cambridge University Press, 1998.

11. Kepler Solves the Planetary Puzzle

Caspar, Max. *Kepler*. New York: Dover, 1993.

Cohen, I. Bernard. *Revolution in Science*. Cambridge: Harvard University Press, 1985.

Crowe, Michael J. *Theories of the World from Antiquity to the Copernican Revolution*. New York: Dover, 1990.

Gillispie, Charles C. *Edge of Objectivity: An Essay in the History of Scientific Ideas*. Princeton: Princeton University Press, 1966.

Huff, Toby E. *The Rise of Early Modern Science: Islam, China, and the West*. Cambridge: Cambridge University Press, 1993.

Koestler, Arthur. *The Sleepwalkers: A History of Man's Changing Vision of the Universe*. London: Hutchinson & Co., 1959.

North, John. *The Norton History of Astronomy and Cosmology*. New York: Norton, 1995.

Ronan, Colin. *The Astronomers*. London: Evans Brothers, 1964.

Stephenson, Bruce. *Kepler's Physical Astronomy*. Princeton: Princeton University Press, 1987.

12. Van Leeuwenhoek Explores the Microcosm

Dobell, Clifford. *Anthony van Leeuwenhoek and his "Little Animals."* New York: Dover, 1960.

Runes, Dagobert D., ed. *A Treasury of World Science*. New York: Philosophical Library, 1962.

Schierbeek, A. *Measuring the Invisible World: The Life and Works of Antoni van Leeuwenhoek FRS*. London and New York: Abelard-Schuman, 1959.

13. Newton: Gravity and Light

Boorstin, Daniel J. *The Discoverers: A History of Man's Search to Know His World and Himself*. New York: Random House, 1983.

Christianson, Gale E. *Isaac Newton and the Scientific Revolution*. Oxford: Oxford University Press, 1996.

Clark, David, and Stephen P. H. Clark. *Newton's Tyranny: The Suppressed Scientific Discoveries of Stephen Gray and John Flamsteed*. New York: W. H. Freeman, 2001.

Koestler, Arthur. *The Sleepwalkers: A History of Man's Changing Vision of the Universe*. London: Hutchinson & Co., 1959.

Maury, Jean-Pierre. *Newton: The Father of Modern Astronomy.* New York: Harry N. Abrams, 1992.

Peterson, Ivars. *Newton's Clock.* New York: W. H. Freeman, 1995.

Thayer, H. S., ed. *Newton's Philosophy of Nature: Selections from His Writings.* New York: Macmillan, 1953.

Westfall, Richard S. *The Life of Isaac Newton.* Cambridge: Cambridge University Press, 1993.

14. A Breath of Fresh Air

Gibbs, F. W. *Joseph Priestley.* New York: Doubleday, 1967.

Kieft, Lester, and Bennett R. Willeford Jr., eds. *Joseph Priestley: Scientist, Theologian, and Metaphysician.* London: Associated University Presses, 1980.

Priestly, Joseph. *Autobiography of Joseph Priestley.* Cranbury, N. J.: Associated University Presses, 1970.

Joseph Priestley: Selections from His Writings. Edited by Ira V. Brown. University Park, Pa. Pennsylvania State University Press, 1962.

Partington, James R. *A Short History of Chemistry.* New York: Dover, 1989.

Schofield, R. E. *The Enlightenment of Joseph Priestley: A Study of His Life and Work from 1733–1773.* University Park, Pa.: Pennsylvania State University Press, 1997.

15. Humphry Davy, Intoxicated with Discovery

Crowther, J. G. *British Scientists of the Nineteenth Century.* London: Routledge & Kegan Paul, 1962.

Fullmer, June Z. *Young Humphry Davy: The Making of an Experimental Chemist.* Philadelphia: American Philosophical Society, 2000.

Hartley, Harold. *Studies in the History of Chemistry.* Oxford: Clarendon Press, 1971.

Knight, David. *Humphry Davy: Science and Power.* Cambridge: Cambridge University Press, 1998.

16. Visionaries of the Computer

Augarten, Stan. *Bit by Bit: An Illustrated History of Computers.* New York: Ticknor & Fields, 1984.

Baum, John. *The Calculating Passion of Ada Byron.* Hamden, Conn.: Archon Books, 1986.

Collier, Bruce, and James MacLachlan. *Charles Babbage and the Engines of Perfection.* Oxford: Oxford University Press, 1998.

Halacy, Dan. *Charles Babbage: Father of the Computer.* New York: Macmillan, 1970.

Hyman, Anthony. *Charles Babbage: Pioneer of the Computer.* New York: Princeton University Press, 1982.

Ifrah, Georges. *The Universal History of Computing: From the Abacus to the Quantum Computer.* New York: John Wiley & Sons, 2001.

Moore, Doris Langley. *Ada, Countess of Lovelace: Byron's Legitimate Daughter.* London: John Murray, 1977.

Stein, Dorothy. *Ada: A Life and a Legacy.* Cambridge: MIT Press, 1985.

Toole, Betty A. *Ada, the Enchantress of Numbers: A Selection from the Letters of Lord Byron's Daughter and Her Description of the First Computer.* Mill Valley, Calif.: Strawberry Press, 1992.

Wooley, Benjamin. *The Bride of Science: Romance, Reason, and Byron's Daughter.* London: Macmillan, 1999.

17. Darwin's Great Truth

Bates, Marston, and Philip S. Humphrey, eds. *The Darwin Reader.* New York: Scribner's, 1956.

Bowlby, John. *Charles Darwin: A New Life.* New York: Norton, 1990.

Darwin, Charles. *The Origin of Species.* New York: Modern Library, 1990.

Dawkins, Richard. *The Blind Watchmaker.* New York: Norton, 1986.

Dennet, Daniel C. *Darwin's Dangerous Idea: Evolution and the Meanings of Life.* New York: Simon and Schuster, 1995.

Eisley, Loren. *Darwin's Century.* New York: Doubleday, 1958.

Mayr, Ernst. *One Long Argument: Charles Darwin and the Genesis of Modern Evolutionary Thought.* Cambridge: Harvard University Press, 1991.

18. A Genius in the Garden

Bronowski, Jacob. *The Ascent of Man.* Boston: Little Brown and Company, 1973.

Cohen, I. Bernard. *Revolution in Science.* Cambridge: Harvard University Press, 1985.

Gribben, John and Mary. *Mendel in 90 Minutes.* London: Constable, 1997.

Henig, Robin Marantz. *The Monk in the Garden: The Lost and Found Genius of Gregor Mendel.* New York: Houghton Mifflin, 2000.

Mendel, Gregor, and Paul C. Mangelsdorf. *Experiments in Plant Hybridization.* Cambridge: Harvard University Press, 1965.

19. Mendeleev Charts the Elements

Ihde, Aaron J. *The Development of Modern Chemistry.* New York: Dover, 1984.

Knedler Jr., John Warren, ed. *Masterworks of Science.* Vol. 2. New York: McGraw-Hill, 1973.

Mendeleeff, Dmitry. *The Principles of Chemistry.* London: Longmans, Green, and Co., 1905 (New York: Kraus Reprint Co., 1969).

Posin, Daniel Q. *Mendleyev: The Story of a Great Scientist.* New York: McGraw-Hill, 1948.

Strathern, Paul. *Mendeleyev's Dream: The Quest for the Elements.* New York: St. Martin's Press, 2001.

20. In the Realm of Radioactivity

Curie, Eve. *Madame Curie.* New York: Doubleday, 1937.

McGrayne, Sharon Bertsch. *Nobel Prize Women in Science: Their Lives, Struggles, and Momentous Discoveries.* New York: Birch Lane Press, 1993.

Pflaum, Rosalyn. *Grand Obsession: Madame Curie and Her World.* New York: Doubleday, 1989.

Quinn, Susan. *Marie Curie: A Life.* New York: Simon & Schuster, 1995.

Reid, Robert. *Marie Curie.* New York: Dutton, 1974.

21. Planck's Quantum Leap

Baggott, Jim. *The Meaning of Quantum Theory.* Oxford: Oxford University Press, 1992.

Gribbin, John. *In Search of Schrödinger's Cat: Quantum Physics and Reality.* New York: Bantam Books, 1984.

Heilbron, J. L. *The Dilemmas of an Upright Man: Max Planck as Spokesman for German Science.* Berkeley: University of California Press, 1986.

Hermann, Armin. *The Genesis of Quantum Theory (1899–1913).* Cambridge: MIT Press, 1971.

Planck, Max. *Scientific Autobiography and Other Papers.* New York: Philosophical Library, 1949.

22. Wired on Wireless

Jolly, W. P. *Marconi.* New York: McGraw-Hill, 1927.

Marconi, Degna. *My Father, Marconi.* Toronto/New York: Guernica Editions, 1996.

Masini, Giancarlo. *Marconi*. New York: Marsilio, 1995.

Reade, Leslie. *Marconi and the Discovery of Wireless*. London: Faber and Faber, 1963.

23. Rutherford Dissects the Atom

Andrade, Edward da Costa. *Rutherford and the Nature of the Atom*. Garden City, N.Y.: Doubleday, 1964.

Eve, A. S. *Rutherford*. Cambridge: Cambridge University Press, 1939.

Feather, Norman. *Lord Rutherford*. London: Priory Press, 1940.

Kelman, Peter, and A. H. Stone. *Ernest Rutherford, Architect of the Atom*. Englewood Cliffs, N.J.: Prentice-Hall, 1968.

Oliphant, Mark. *Rutherford: Recollections of the Cambridge Days*. Amsterdam: Elsevier, 1972.

Wilson, David. *Rutherford: Simple Genius*. Cambridge: MIT Press, 1983.

24. Einstein: Matter, Energy, Space, and Time

Bartusiak, Marcia. *Einstein's Unfinished Symphony: Listening to the Sounds of Space-Time*. Washington, D.C.: Joseph Henry, 2000.

Brian, Denis. *Einstein: A Life*. New York: John Wiley and Sons, 1996.

Calaprice, Alice. *The Quotable Einstein*. Princeton: Princeton University Press, 1996.

De Broglie, Louis, et al. *Einstein*. New York: Peebles Press, 1979.

Dukas, Helen, and Banesh Hoffman. *Albert Einstein: The Human Side*. Princeton: Princeton University Press, 1979.

Folsing, Albrecht. *Albert Einstein: A Biography*. New York: Viking, 1997.

Goldsmith, Donald, and Robert Libbon. *The Ultimate Einstein*. New York: Byron Press, 1997.

Goldsmith, Maurice, et al., eds. *Einstein: The First Hundred Years*. Oxford: Pergamon, 1980.

Overbye, Dennis. *Einstein in Love: A Scientific Romance*. New York: Viking, 2000.

Pais, Abraham. *"Subtle is the Lord . . .": The Science and the Life of Albert Einstein*. Oxford: Clarendon Press, 1982.

25. Wegener Sets the Continents Adrift

Cohen, I. Bernard. *Revolution in Science*. Cambridge: Harvard University Press, 1985.

Continental Drift Animation can be found online at http://www.ucmp .berkeley.edu/geology/anim1.html

Hallam, Anthony. *A Revolution in the Earth Sciences: From Continental Drift to Plate Tectonics.* Oxford: Clarendon Press, 1973.

Oreskes, Naomi. *The Rejection of Continental Drift: Theory and Method in American Earth Sciences.* Oxford: Oxford University Press, 1999.

Schwarzbach, Martin. *Alfred Wegener: The Father of Continental Drift.* Madison, Wis.: Science Tech, Inc. 1986.

Wegener, Alfred L., and Kurt Wegener. *The Origin of Continents and Oceans.* New York: Dover, 1966.

26. Hubble's Expanding Universe

Christianson, Gale E. *Edwin Hubble: Mariner of the Nebulae.* New York: Farrar, Straus and Giroux, 1995.

Ferris, Timothy. *Coming of Age in the Milky Way.* New York: Morrow, 1988.

Ferris, Timothy. *The Whole Shebang: A State-of-the-Universe(s) Report.* London: Orion Books, 1998.

Hawley, John F., and Katherine A. Holcomb. *Foundations of Modern Cosmology.* Oxford: Oxford University Press, 1998.

Hubble, Edwin. *The Nature of Science and Other Essays.* San Marino: Huntington Library, 1954.

Overbye, Dennis. *Lonely Hearts of the Cosmos: The Story of the Quest for the Secret of the Universe.* New York: Harper Collins, 1991.

Smith, Robert W. *The Expanding Universe: Astronomy's "Great Debate," 1900–1931.* Cambridge: Cambridge University Press, 1982.

27. Out of Africa

Dart, Raymond A., with Dennis Craig. *Adventures with the Missing Link.* London: Hamish Hamilton, 1959.

Johanson, Donald, and Blake Edgar. *From Lucy to Language.* New York: Simon and Schuster, 1996.

Leakey, L. S. B., and Jack and Stephanie Prost, eds. *Adam, or Ape: A Sourcebook of Discoveries About Early Man.* Cambridge, Mass.: Schenkman, 1971.

Leakey, Richard, and Roger Lewin. *Origins Reconsidered: In Search of What Makes Us Human.* New York: Doubleday, 1992.

Lewin, Roger. *Bones of Contention: Controversies in the Search for Human Origins.* New York: Simon & Schuster, 1987.

Tattersall, Ian. *The Fossil Trail: How We Know What We Think We Know about Human Evolution*. Oxford: Oxford University Press, 1995.

Wheelhouse, Frances, and Kathaleen Smithford. *Dart: Scientist and Man of Grit*. Hornsby, Australia: Transpareon Press, 2001.

28. Fermi and the Fire of the Gods

Cooper, Dan. *Enrico Fermi and the Revolutions of Modern Physics*. Oxford: Oxford University Press, 1999.

Fermi, Laura. *Atoms in the Family: My Life with Enrico Fermi*. Chicago: University of Chicago Press, 1954.

Rhodes, Richard. *The Making of the Atomic Bomb*. New York: Simon & Schuster, 1986.

Segrè, Emilio. *Enrico Fermi: Physicist*. Chicago: University of Chicago Press, 1970.

29. McClintock's Chromosomes

Hammond, Allen L., ed. *A Passion to Know: 20 Profiles in Science*. New York: Scribner's, 1984.

Heiligman, Deborah. *Barbara McClintock: Alone in Her Field*. New York: W. H. Freeman, 1994.

Keller, Evelyn Fox. *A Feeling for the Organism: The Life and Work of Barbara McClintock*. New York: Freeman, 1983.

McGrayne, Sharon B. *Nobel Prize Women in Science*. New York: Birch Lane, 1993.

Peters, James A., ed. *Classic Papers in Genetics*. Englewood Cliffs, N.J.: Prentice-Hall, 1959.

Shepherd, Linda Jean. *Lifting the Veil: The Feminine Face of Science*. Boston: Shambala, 1993.

Veglahn, Nancy J. *Women Scientists*. New York: Facts on File, 1991.

Voytas, Daniel F. "Retroelements in Genome Organization." *Science* 274, no. 5288 (1996): 737–738.

30. A Bit of Genius

Horgan, John. "Claude E. Shannon: Unicyclist, juggler, and father of information theory." *Scientific American* 262 (1): 22–22B.

Liversidge, Anthony. "Claude Shannon." *OMNI*, August 1987, 61.

Shannon, Claude E., and Warren Weaver. *The Mathematical Theory of Communication*. Urbana, Ill.: University of Illinois Press, 1999.

Sloane, N. J. A., and Aaron D. Wyner. *Claude Elwood Shannon: Collected Papers*. Piscataway, N.J.: IEEE Press, 1993.

31. The Dynamic Duo of DNA

Crick, Francis. "On the Genetic Code." In *Nobel Lectures in Molecular Biology: 1933–1975*. New York: Elsevier, 1977.

Drlica, Karl. *Understanding DNA and Gene Cloning: A Guide for the Curious*, 3rd edition. New York: John Wiley & Sons, 1996.

Gribben, John. *In Search of the Double Helix: Quantum Physics and Life*. New York: McGraw-Hill, 1985.

Hoagland, Mahlon. *Discovery: The Search for DNA's Secrets*. Boston: Houghton Mifflin, 1981.

Judson, Horace F. *The Eighth Day of Creation: Makers of the Revolution in Biology*. New York: Simon and Schuster, 1979.

Olby, Robert. *The Path to the Double Helix*. Seattle: University of Washington Press, 1974.

Sayre, Anne. *Rosalind Franklin and DNA*. New York: Norton, 1975.

Watson, James D. *The Double Helix: A Personal Account of the Discovery of the Structure of DNA*. New York: Penguin, 1969.

32. Echoes of Creation

Bernstein, Jeremy. *Three Degrees Above Zero: Bell Labs in the Information Age*. New York: Scribner's, 1984.

Chown, Marcus. *Afterglow of Creation: From the Fireball to the Discovery of Cosmic Ripples*. Sausalito, Cal.: University Science Books, 1996.

Fox, Karen C. *The Big Bang: What It Is, Where It Came From, and Why It Works*. New York: John Wiley & Sons, 2002.

Hawley, John F., and Katherine A. Holcomb. *Foundations of Modern Cosmology*. Oxford: Oxford University Press, 1998.

Mather, John C., and John Boslough. *The Very First Light: The True Inside Story of the Scientific Journey Back to the Dawn of the Universe*. New York: Basic Books, 1996.

Partridge, R. B. *3 K: The Cosmic Microwave Background Radiation*. Cambridge: Cambridge University Press, 1995.

Silk, Joseph. *The Big Bang, 3d Edition*. New York: Freeman, 2001.

Smoot, George, and Keay Davidson. *Wrinkles in Time*. New York: Morrow, 1993.

Weinberg, Steven. *The First Three Minutes: A Modern View of the Origin of the Universe*. New York: Basic Books, 1977.

33. We Are Not What We Seem

Margulis, Lynn. *Diversity of Life: The Five Kingdoms*. Hillside, N.J.: Enslow Publishers, 1992.

Margulis, Lynn. *Symbiotic Planet: A New Look at Evolution*. New York: Basic Books, 1998.

Margulis, Lynn, and Dorion Sagan. *Microcosmos: Four Billion Years of Microbial Evolution*. New York: Summit Books, 1986.

Margulis, Lynn, Dorian Sagan, and Ernst Mayr. *Acquiring Genomes: A Theory of the Origins of Species*. New York: Basic Books, 1992.

Sapp, Jan. *Evolution by Association: A History of Symbiosis*. Oxford: Oxford University Press, 1994.

34. Planetary Pioneers

Boss, Alan. *Looking for Earths: The Race to Find New Solar Systems*. New York: John Wiley & Sons, 1998.

Croswell, Ken. *Planet Quest: The Epic Discovery of Alien Solar Systems*. New York: Free Press, 1997.

Goldsmith, Donald. *Worlds Unnumbered: The Search for Extrasolar Planets*. Sausalito, Cal.: University Science Books, 1997.

Halpern, Paul. *The Quest for Alien Planets: Exploring Worlds Outside the Solar System*. New York: Plenum, 1997.

Lemonick, Michael. *Other Worlds: The Search for Life in the Universe*. New York: Simon and Schuster, 1998.

35. After Dolly, Life Will Never Be the Same

Keller, Evelyn Fox. *The Century of the Gene*. Cambridge: Harvard University Press, 2000.

Kolata, Gina B. *Clone: The Road to Dolly, and the Path Ahead*. New York: Morrow, 1998.

Pence, Gregory E. *Who's Afraid of Human Cloning?* Lanham, Md.: Rowman & Littlefield Publishers, 1998.

Pence, Gregory E., ed. *Flesh of My Flesh: The Ethics of Cloning Humans*. Lanham, Md.: Rowman & Littlefield Publishers, 1998.

Wilmut, Ian, Keith Campbell, and Colin Tudge. *The Second Creation: Dolly and the Age of Biological Control*. New York: Farrar, Strauss and Giroux, 2000.

Winters, Paul A., ed. *Cloning*. San Diego: Greenhaven Press, 1998.

Index

232   INDEX